International and European Environmental Policy Series

Edited by
R. Andreas Kraemer
Sascha Müller-Kraenner

Eleftheria Kampa • Wenke Hansen

Heavily Modified Water Bodies

Synthesis of 34 Case Studies in Europe

With 19 Figures and 49 Tables

Springer

Authors

Eleftheria Kampa
Wenke Hansen

Ecologic Institute for International and European
Environmental Policy
Pfalzburger Straße 43/44
10717 Berlin, Germany
Tel: 0049 +30 86880-0
Fax 0049 +30 86880-100
E-mail: office@ecologic.de

Cover photo: Overview of the Haringvliet dam and sluices
(Source: Case Study on the Haringvliet Estuary in the Netherlands)

Library of Congress Control Number: 2004106674

ISBN 3-540-21085-7 Springer Berlin Heidelberg New York

Springer is a part of Springer Science+Business Media
springeronline.com

Printed in Germany

Cover design: E. Kirchner
Production: A. Oelschläger
Typesetting: Büro Stasch
Printing: Mercedes-Druck

Printed on acid-free paper 30/2132/AO 5 4 3 2 1 0

Preface

Disclaimer

The work reported in this book was carried out prior to the publication of the final guidance document on Heavily Modified and Artificial Water Bodies of the Water Framework Directive Common Implementation Strategy. The European Commission, the project donors, the project managers, the Working Group on Heavily Modified Water Bodies of the Common Implementation Strategy and the authors do not necessarily endorse all approaches that were taken in the case studies that are reported.

About Ecologic

Ecologic, Institute for International and European Environmental Policy, is a private not-for-profit think tank for applied environmental research, policy analysis and consultancy with offices in Berlin and Brussels. An independent, non-partisan body, Ecologic is dedicated to bringing fresh ideas to environmental policies and sustainable development. Ecologic's work programme focuses on obtaining practical results. It covers the entire spectrum of environmental issues and includes the integration of environmental concerns into other policy fields.

Ecologic is the German partner in the network of Institutes for European Environmental Policy. Founded in 1995, Ecologic has a multi-disciplinary staff that includes lawyers, negotiators, social and natural scientists, economists, technologists, and planners. With a focus on Europe and North America, Ecologic's activities cover a growing geographical area and involve policy-makers, business leaders, technical experts and representatives of civil society organisations.

Ecologic is officially recognised as being in the public interest. Donations to Ecologic are tax-deductible.

Acknowledgements

This book is based on the results of thirty-four case studies that were carried out in the context of the Working Group on Heavily Modified Water Bodies (HMWB) of the Common Implementation Strategy (CIS) of the EU Water Framework Directive (2000/60/EC) (details on the CIS are provided in the introductory chapter). In a European Synthesis Project on HMWB carried out in the context of the CIS, Ecologic (Institute for International and European Environmental Policy) was commissioned by the European Commission and the Joint Chair (UK/D) of the CIS Working Group on HMWB to comprehensively analyse and summarise the main case study results. The authors of this book are particularly grateful to the following organisations that funded the European Synthesis Project on HMWB:

- Scotland and Northern Ireland Forum for Environmental Research (SNIFFER);
- Environment Agency of England and Wales;
- German Federal Environment Agency (Umweltbundesamt); and
- Directorate-General for the Environment of the European Commission.

The authors would particularly like to thank SNIFFER and the Environment Agency of England and Wales for providing additional funds for the publication of this book.

We would like to address special thanks to the Joint Chair of the CIS Working Group (2.2) on Heavily Modified Water Bodies: Dr. David Forrow (Environment Agency of England and Wales, UK), Dr. Ulrich Irmer (Umweltbundesamt, D), Martin Marsden (Scottish Environment Protection Agency,[1] UK), and Dr. Bettina Rechenberg (Umweltbundesamt, D). Their support, comments and advice have largely contributed to the writing of the synthesis document on HMWB, which has been the basis for this book.

The thirty-four case study reports, which provided the empirical material for this book, were contributed by the different national teams participating in the CIS Working Group on HMWB. We would like to use this opportunity to thank all *contributing authors* of this book:

- The authors of the case study reports (listed in Table 13.2 of Annex I, p. 180); and
- The members of the Working Group 2.2 on HMWB (listed in Table 13.1 of Annex I, p. 177).

All written contributions, comments and advice made the writing of this book possible and must be duly acknowledged. If more information on specific case studies is sought, the persons in charge of carrying out these case studies can be contacted directly (contacts are listed in Table 13.3 of Annex I, p. 181). We did our best to interpret and resume the HMWB case study reports correctly. Any mistakes remain of our own.

Finally, we would like to thank Christine Laskov for her invaluable contribution to the analysis of the first drafts of the case study reports and in contributing to an earlier draft of this document, R. Andreas Kraemer (director of the Institute Ecologic) for his conceptual contribution towards the HMWB Synthesis Project, Maic Verbuecheln for his help in coordinating the collection of the case study reports as well as Martin Obermaier, Thomas Reidel and Antje Vorwerk for their formatting skills.

[1] The Scottish Environment Protection Agency (SEPA) is one of five organisations – members of the Scotland and Northern Ireland Forum for Environmental Research (SNIFFER). SNIFFER identifies and manages environmental research on behalf of its members.

Contents

List of Figures

List of Tables

List of Boxes

Abbreviations

A	Austria
Art.	Article
a.s.l.	Above sea level
AWB	Artificial Water Body/Bodies
B	Belgium
BMWP	Biological Monitoring Working Party, UK
CIS	Common Implementation Strategy
D	Germany
DHRAM	Dundee Hydrological Regime Alteration Method, UK
E	Spain
EC	European Commission
EQR	Ecological Quality Ratio
ES	Ecological Status
EU	European Union
F	France
FDMS	Flood Defence Management System, UK
GEP	Good Ecological Potential
GES	Good Ecological Status
GIS	Geographical Information Systems
GQA	General Quality Assessment, UK
GR	Greece
HES	High Ecological Status
HMS	Habitat Modification Score, UK
HMWB	Heavily Modified Water Body/Bodies
HP	Hydropower
IBI	Index of Biotic Integrity
km	Kilometre
km^2	Square-kilometres
LAWA	Länderarbeitsgemeinschaft Wasser (Working Group of the Federal States on Water Problems), Germany
MEP	Maximum Ecological Potential

MS	Member State
NL	The Netherlands
NO	Norway
PA	Physical Alteration
POM	Programme Of Measures
QBR	Q = Qualitat (Qualit y); B = Bosc (Forest); R = Ribera (Riparian)
RBMP	River Basin Management Plan
RHS	River Habitat Survey, UK
RIVPACS	River Invertebrate Prediction and Classification System, UK
S	Sweden
SF	Finland
SPA	Special Protection Area
SSSI	Site of Special Scientific Interest, UK
UK	United Kingdom
UK-EW	UK-England and Wales
UK-NI	UK-Northern Ireland
UK-S	UK-Scotland
WB	Water Body
WFD	Water Framework Directive
WG	Working Group

Chapter 1

Introduction

This book is based on the results of the European Synthesis Project of thirty-four case studies on the identification and designation of Heavily Modified Water Bodies (HMWB) in the context of the EU Water Framework Directive (2000/60/EC)[1]. The European Synthesis Project on HMWB was part of the activities of the Working Group on HMWB of the Common Implementation Strategy (CIS) of the Water Framework Directive (WFD). To put this book into context, the following section briefly reviews the policy background to the European Synthesis Project on HMWB.

1.1 Policy Background

In accordance with Article 4(3) of the EU WFD, the Directive allows EU Member States to identify surface water bodies that have been physically altered by human activity as "heavily modified" under specific circumstances. Water bodies can be designated as "heavily modified" and Good Ecological Potential set as the environmental objective if

- The specified uses of such water bodies (e.g. navigation, hydropower, water supply or flood defense) or the "wider environment" would be significantly affected by the restoration measures required to achieve Good Ecological Status; and
- No other technically feasible, cost-effective and better environmental options exist that would maintain the benefits of the modified characteristics of the water body.

Under certain conditions, the WFD also permits EU Member States to identify and designate Artificial Water Bodies (AWB) in accordance with Article 2(8) of the EU WFD: "*artificial water body means a body of surface water created by human activity*".

1.1.1 The Common Implementation Strategy of the Water Framework Directive

The need for establishing a Common Implementation Strategy (CIS) for the Water Framework Directive was identified at an informal meeting of the EU Water Directors and the Norwegian Water Director in Paris on 23–24 October 2000.

[1] Directive 2000/60/EC of the European Parliament and of the Council of 23 October 2000, establishing a framework for community action in the field of water policy, Official Journal 22 December 2000 L 327/1, European Commission, Brussels.

According to the strategic document developed for the CIS,[2] the implementation of the Water Framework Directive raises challenges, which are widely shared by Member States. The Directive sets a demanding timetable for EU Member States, especially during the nine preparatory years until the publication of the first river basin management plans in 2009. The aim of the CIS is therefore to contribute, as far as possible, to a timely, coherent and harmonious implementation.

The CIS focuses on methodological questions related to a common understanding of technical and scientific implications of the WFD. In this respect, the main elements of the CIS as identified at the meeting of the Water Directors on 23–24 October 2000 (in Paris) included the need to establish working groups that would develop guidance documents on key aspects of the WFD, illustrating several technical and scientific issues. The guidance documents of the CIS have an informal, legally non-binding character and are placed at the disposal of EU Member States who wish to use them on a voluntary basis.

The CIS additionally recognises that each Member State will undoubtedly face specific questions and challenges related to national, regional and/or local situations. The strategy was thereby developed in recognition and respect for these specific situations.[3]

Regarding the aspects of work of the CIS working groups, a limited number of key activities have been identified on which Member States and the European Commission have jointly worked. These activities are illustrated in Fig. 1.1, which is a schematic representation of the overall structure of the CIS. According to this modular structure, four key activities are identified: (1) information sharing, (2) developing guidance on technical issues, (3) information and data management and (4) application, testing and validation. These activities were set as priorities for the first important phase of the CIS. The first three activities have a more horizontal character and aim at developing a common understanding of the implementation of the WFD. Their results need to be integrated and made operational in the river basin management plans. The fourth activity has a more vertical character and includes the practical integration of the horizontal activities through testing in pilot river basins.[4]

In this context, working groups have been established for the different activities and projects. Working groups have generally been chaired by one or two lead countries or the European Commission with participants from interested EU Member States, future and new EU Member States and stakeholders. The established working groups correspond to the projects of the second activity (development of guidance) as well as to the third activity (development of a shared Geographical Information System). The work of the different working groups has been very strongly linked. The overall co-ordination of the different working groups and activities has been ensured by the Strategic Co-ordination Group of the WFD CIS (chaired by the European Commission). This coordination aimed at avoiding duplication of work and ensuring the necessary exchange of information.

The working groups have therefore produced guidance documents in a pragmatic way based on existing practices in Member States. In this context, the guidance documents aim to be practical, operational as well as policy and implementation oriented.

[2] Strategic document: Common Strategy on the Implementation of the Water Framework Directive. 2 May 2001. Available online: http://forum.europa.eu.int/Public/irc/env/wfd/home.

[3] Strategic Document: Common Strategy on the Implementation of the Water Framework Directive. 2 May 2001. Available online: http://forum.europa.eu.int/Public/irc/env/wfd/home.

[4] Ibid.

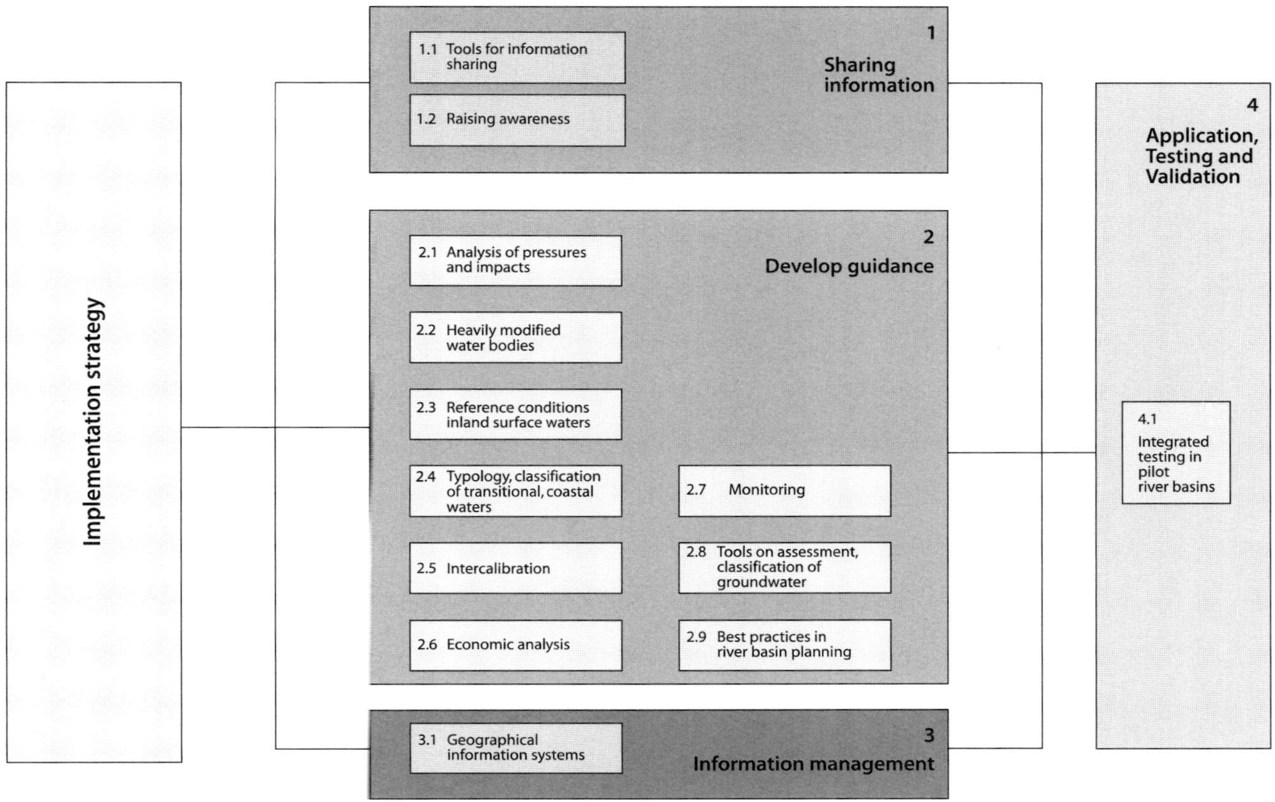

Fig. 1.1. Scheme of the activities and working groups of the Common Implementation Strategy. *Source:* Strategic Document. Common Strategy on the Implementation of the Water Framework Directive. 2 May 2001

Most of the CIS guidance documents had been agreed and made available to the public by mid-2003. The complete guidance documents thereby provided the basis for the integrated testing in pilot river basins of the CIS (fourth activity of the CIS presented in Fig. 1.1). The aim of the latter is to test these documents in selected pilot river basins in order to identify the technical and management problems that may arise in real cases of the implementation of the Water Framework Directive.

A follow-up of the Common Implementation Strategy (CIS) was also agreed at the meeting of the Water Directors in Copenhagen on 21–22 November 2002. The Water Directors' decision was to establish four new cluster working groups on the following issues: Ecological Status (Cluster Working Group 2.A), Integrated River Basin Management (Cluster Working Group 2.B), Groundwater (Cluster Working Group 2.C), and Reporting (Cluster Working Group 2.D). Therefore, the different working groups of the first phase of the CIS (Fig. 1.1) were dissolved and their activities were clustered into the new working groups on cross-cutting issues of the WFD.

1.1.2 CIS Working Group on Heavily Modified Water Bodies

As part of the WFD Common Implementation Strategy (CIS), a working group was established to develop guidance on the issue of Heavily Modified and Artificial Water Bodies (HMWB and AWB) identification and designation. This CIS Working Group on Heavily Modified Water Bodies (see Project 2.2 in Fig. 1.1) was jointly chaired by the United Kingdom and Germany. The Working Group (WG) was officially comprised of representatives from twelve EU Member States[5], Norway, some new and future EU Member States[6], and a number of stakeholders[7] (see list in Table 13.1 of Annex I, p. 177).

The CIS Working Group (WG) on HMWB decided to base the WG discussions and the development of the CIS guidance document on HMWB and AWB around a number of case studies conducted in the countries that participated in the Working Group. The case studies were to test a number of supporting guidance papers developed by the joint chair of the WG, which were reviewed and agreed by the WG members. This approach was unique in the Common Implementation Strategy, as the HMWB Working Group was the only CIS WG to initiate case studies in order to develop its final guidance document. This empirical approach was considered necessary because there was little, if any, experience of HMWB and AWB identification and designation in the Member States.

In this context, a number of distinct "subprojects" were developed by the HMWB WG in the framework of the Common Implementation Strategy:

[5] Austria, Belgium, Denmark, Spain, France, Germany, Greece, the Netherlands, Portugal, Sweden, Finland, and the UK.

[6] Hungary, Poland and Slovenia. Other future and new EU Member States were also members of the WG but did not attend the WG meetings.

[7] European Environmental Bureau (EEB), European Union of National Associations of Water Suppliers and Waste Water Services (EUREAU), Union of the Electricity Industry (Eurelectric) and World Wildlife Fund (WWF).

- Production of twelve guidance papers by the joint chair of the HMWB WG (as support material for conducting the case studies on Heavily Modified Water Bodies). The guidance papers were discussed at several Working Group meetings;
- Thirty-four case study projects carried out in the EU Member States and Norway that tested these "guidance papers";
- *A synthesis of the thirty-four case study project reports (European Synthesis Project, which is the basis of this book)*;
- Production of the CIS guidance document on HMWB and AWB;
- Production of a policy summary of the CIS guidance document on HMWB and AWB; and
- Production of a toolbox supporting the CIS guidance document on HMWB and AWB.

Ecologic (Institute for International and European Environmental Policy) was commissioned to provide conceptual and organisational support to the CIS Working Group on HMWB since the beginning of its activities. The support involved several tasks, which were mainly the following: production of the synthesis document of the thirty-four HMWB case studies, development of the CIS guidance document, policy summary and toolbox together with the joint chair and the members of the Working Group, conceptual and technical organisation of the Working Group meetings as well as facilitation of the flow of information and communication.

The distinct "subprojects" of the HMWB Working Group are described in more detail in the following paragraphs.

Production of Twelve Guidance Papers

The joint chair of the HMWB WG produced twelve guidance papers covering the key aspects of the HMWB and AWB identification and designation process. Four meetings were organised involving the Working Group members and the European Commission to discuss and agree these guidance papers and to exchange experiences. The meetings were held on 12 April 2000, 10 October 2000, 4 September 2001 and 18–19 June 2002 in Brussels. The guidance papers were developed to guide a series of case studies on Heavily Modified Water Bodies, which tested the approaches of the guidance papers. Together with the case study reports, the twelve guidance papers served partly as the basis for the production of the CIS guidance document on HMWB and AWB.

Case Studies

The twelve guidance papers on the identification and designation of Heavily Modified Water Bodies (HMWB) were tested in thirty-four case studies. The latter were carried out in eleven countries[8] participating in the WG on HMWB. In the case studies, ecological reference conditions (Maximum Ecological Potential) and objectives (Good Ecological Potential) for HMWB were also defined for the water bodies of the case study areas, as much as possible.

[8] Norway (NO) and ten EU Member States: Austria (A), Belgium (B), Spain (E), France (F), Germany (D), Greece (GR), Netherlands (NL), Sweden (S), Finland (SF) and UK.

The case studies focused on the main specified water uses (navigation, flood/coastal protection, hydropower generation, agriculture, forestry, urbanisation, recreation, and water supply) that result in physical alterations. Member States participating in the WG on HMWB were given the flexibility to choose their own case study location(s). The sites that were chosen tended to be those for which Member States had available data. The case study projects started in October 2000 and were finalised in June 2002. The WG on HMWB did not necessarily endorse the approach taken by any individual case study. It should be noted that the case studies do not strictly follow the approach of the CIS guidance document on HMWB and AWB, since most of them were completed before the publication of the final approved version of the guidance document (published in January 2003).

Based on the main water uses within the case studies, two case study subgroups were also established on navigation and on hydropower (see Table 14.1 in Annex II). The HMWB Working Group members and/or case study contractors exchanged their experiences during their work in extra subgroup meetings and via email discussions.

European Synthesis Project

The European Synthesis Project on HMWB performed an analysis and synthesis of the approaches taken in the individual HMWB case studies, identifying commonalities and differences. The analysis started in February 2002, and a first draft synthesis document was distributed to the HMWB WG by the end of April 2002. The first draft of the synthesis document formed the basis for the production of the HMWB and AWB guidance document (WFD CIS Guidance Document No. 4 2003) as well as the toolbox supporting the guidance document (WFD CIS Working Group 2.2 on HMWB 2003a).

This book is based on the final results of the European Synthesis Project of the HMWB case studies, being an evaluation of all thirty-four HMWB case study reports. For the production of the final synthesis, the terminology used in the first draft of the synthesis document (dated April 2002) was adjusted to the glossary agreed in the final HMWB and AWB guidance document (WFD CIS Guidance Document No. 4 2003).

The CIS Guidance Document on HMWB and AWB

Based on the draft synthesis document and on the twelve guidance papers of the HMWB CIS Working Group, a first draft guidance document on the identification and designation of HMWB and AWB[9] was produced on 27 May 2002. A workshop was held on the 30–31 May 2002 in Berlin for Working Group members, case study contractors, and members of other CIS working groups[10] to discuss a number of outstanding is-

[9] Guidance document on identification and designation of Heavily Modified Water Bodies, first draft, CIS Working Group 2.2 on Heavily Modified Water Bodies, 27 May 2002.

[10] There are important links between the HMWB and other CIS working groups. These included the Working Group 2.1 on pressures and impacts, Working Group 2.3 on freshwater reference conditions, Working Group 2.4 on coastal water typology, reference and classification, Working Group 2.5 on intercalibration, Working Group 2.6 on economic analysis, Working Group 2.7 on monitoring, Working Group 2.9 on best practices in river basin planning and Working Group 3.0 on Geographical Information Systems (GIS).

sues with regard to the draft guidance on HMWB and AWB. The discussions during the workshop served as a basis for the revision of the draft guidance document. A second draft[11] was discussed at the last HMWB WG meeting in June 2002. A third draft[12] was produced and circulated to the WG for comments in August 2002. A final version of the HMWB and AWB guidance document[13] was produced and submitted to the WFD CIS strategic co-ordination group meeting on 30 September 2002. It was then revised and presented to the WFD CIS strategic co-ordination group meeting on 7–8 November 2002. The final version of the HMWB and AWB guidance document was agreed at the Water Directors' meeting on 21–22 November 2002 and published on 14 January 2003 (WFD CIS Guidance Document No. 4 2003). The HMWB and AWB guidance document represents a consensus view of the EU Member States represented in the HMWB WG, Norway, the participating future and new EU Member States and stakeholders.

The Policy Summary

The policy summary is an executive summary of the HMWB and AWB guidance document addressed to the Water Directors. The document summarises the main issues of the HMWB and AWB designation process and is derived directly from the HMWB and AWB guidance document. It was presented and agreed at the Water Directors' meeting together with the guidance document on 21–22 November 2002.[14]

The Toolbox

To support the HMWB and AWB guidance document, a toolbox was produced, illustrating the different steps of the HMWB and AWB designation process with practical examples from the HMWB case studies and the synthesis document. The HMWB Working Group members also provided additional examples that helped illustrate certain steps of the guidance document. A first draft was produced for the HMWB WG meeting in June 2002. A second draft was sent out for comments in October 2002 and a final toolbox was issued in January 2003 (WFD CIS Working Group 2.2 on HMWB 2003a). The toolbox does not constitute part of the HMWB and AWB guidance document and was hence not subject to the agreement of the HMWB Working Group.

Links to other CIS Working Groups

It is important to consider this book as well as the HMWB and AWB guidance document (WFD CIS Guidance Document No. 4 2003) and toolbox (WFD CIS Working

[11] Guidance document on identification and designation of (Artificial and) Heavily Modified Water Bodies, second draft, CIS Working Group 2.2 on Heavily Modified Water Bodies, 15 June 2002. Directly after the WG meeting in June, a second draft dated 20 June was sent to the WG, including a different version of Chap. 6.

[12] Guidance Document on identification and designation of Artificial and Heavily Modified Water Bodies, third draft, CIS Working Group 2.2 on Heavily Modified Water Bodies, 2 August 2002.

[13] Guidance document on identification and designation of Artificial and Heavily Modified Water Bodies, final draft, CIS Working Group 2.2 on Heavily Modified Water Bodies, 13 September 2002.

[14] WFD CIS Working Group 2.2 on HMWB (Jan 2003b).

Group 2.2 on HMWB 2003a) in the context of the guidance documents produced by other working groups of the WFD CIS. Important links exist between the HMWB Working Group and other CIS working groups such as the CIS Working Group 2.1 on pressures and impacts (IMPRESS), Working Group 2.3 on freshwater reference conditions (REFCOND), Working Group 2.4 on transitional and coastal water typology, reference and classification (COAST), Working Group 2.5 on intercalibration, Working Group 2.6 on economic analysis (WATECO), Working Group 2.7 on monitoring, Working Group 2.9 on the best practices in river basin planning and Working Group 3.0 on Geographical Information Systems (GIS).

As far as possible, links, feedback and common areas of work have been identified throughout this book. The HMWB and AWB guidance document (WFD CIS Guidance Document No. 4 2003) provides more details on the links to other CIS workinggroups.

1.2 Objectives

As part of the activities of the Working Group on HMWB of the WFD Common Implementation Strategy, this book aims to identify commonalities and differences in the approaches of thirty-four case studies on the process of identification and designation of HMWB and AWB (European Synthesis Project on HMWB). The analysis of the thirty-four case studies aims to provide examples of how the HMWB designation process was approached and to illustrate the variance of country-specific approaches on the technical implications of the WFD for HMWB and AWB. The initial policy objective of the HMWB European Synthesis Project was to synthesise the experience of the HMWB case studies and to form an empirical basis for the production of the CIS guidance on HMWB and AWB. This policy objective was fulfilled with the publication of the guidance document on HMWB and AWB in January 2003 as part of the Common Implementation Strategy of the EU Water Framework Directive.

In the long term, this book aims to serve as a first reference and empirical basis for future research and management of Heavily Modified and Artificial Water Bodies. The implementation of the WFD continues after the publication of the CIS guidance documents through integrated testing in pilot river basins throughout Europe and other nationally-based research projects. This book therefore aims to assist authorities, researchers and consultants involved in the implementation of the WFD and specifically with the identification and designation of HMWB and AWB as described in the respective CIS guidance document.

1.3 Methodology

An in-depth comparative analysis of the thirty-four HMWB case studies was carried out starting in February 2002. The results and main conclusions of the case studies have been synthesised with focus on the steps of the identification and designation process for HMWB and AWB (see Fig. 2.1).

The comparative evaluation of the HMWB case studies has been assisted by the fact that their authors followed a standard reporting outline agreed within the HMWB Working Group. This reporting outline included information on

- The case study area;
- Physical alterations;
- Ecological status;
- (Provisional) identification and designation of Heavily Modified Water Bodies;
- Definition of Maximum Ecological Potential (MEP);
- Definition of Good Ecological Potential (GEP);
- Conclusions, options and recommendations.

Evaluation tables were developed to extract and compile the main results, conclusions and encountered problems from the thirty-four case studies. The information provided by the case studies was analysed and synthesised on both a case-by-case basis and according to the different issues of the HMWB identification and designation process.

A first draft synthesis document was prepared by the end of April 2002. For the further development and revision of the draft synthesis document which was ready in April 2002, an interactive approach was followed by distributing the document to the HMWB WG and the case study authors for comments. The discussions during the workshop on HMWB, held on 30–31 May 2002 in Berlin, as well as written feedback and comments served as a basis for the revision and further development of the draft synthesis document. The final version of the synthesis document of the HMWB case studies was prepared by March 2003. This book is based on the final synthesis but has been enriched with reference to the overall CIS process as well as the CIS guidance document on HMWB and AWB in order to be of use to a broader audience.

1.4 Scope and Structure

This book concentrates on the identification and designation of water bodies as heavily modified according to the requirements of the WFD. The issue of differentiation of Artificial Water Bodies (AWB) from Heavily Modified Water Bodies (HMWB) is only briefly dealt within Sect. 4.4 of the book, summarising the main findings of the case studies. This is because AWB were not extensively dealt with in the first drafts of the guidance document of the HMWB WG. However, the final guidance document on HMWB and AWB deals much more explicitly with AWB.

In this book, reference is also made to several technical and scientific issues of the WFD such as "water body identification", the concept of "pressures and impacts" and "reference conditions". Nevertheless, extensive discussions on these issues are not within the scope of this document. The reader should refer to the related CIS guidance documents prepared by the European Commission and the respective CIS working groups on these issues for more explicit guidance and definitions (see Sect. 1.1.1

of this book).[15] Work on the HMWB case studies actually started earlier than most of other CIS projects. Consequently, there were no defined approaches to some of the technical terms of the WFD which are precursors to the identification of HMWB in the river basin management planning process of the WFD. An example of such a requirement is the development of approaches to identify typologies and water bodies. This difference in the timing of the CIS activities led to numerous assumptions and simplifications in carrying out the HMWB case studies.

For the purpose of reporting the results and conclusions from the HMWB case studies, the authors summarised the approaches mostly according to the different water categories (rivers, lakes, transitional and coastal waters) and where possible, according to different categories of specified water uses.

The structure of the book, which follows the structure of the case study reporting outline to a great extent, is as follows:

Chapter 1 is an introduction to the policy background, the objectives, methodology and structure. Chapter 2 is dedicated to the final CIS guidance document on HMWB and AWB, illustrating the steps of the approved process for the identification and designation of HMWB and AWB. For the reader who is not familiar with the CIS guidance document on HMWB and AWB, it is recommended that this chapter is read first in order to understand later reference in the book to steps of the identification and designation process. Chapter 3 introduces the thirty-four HMWB case studies. Chapters 4 through 10 are dedicated to the synthesis of results and conclusions of the thirty-four case studies on the distinct issues and steps of the identification and designation process: Chapter 4 on the identification of water bodies, Chap. 5 on the description of specified uses, physical alterations and impacts on hydromorphology, Chap. 6 on the assessment of ecological status (and failure to achieve it), Chap. 7 on the provisional identification of HMWB, Chap. 8 on the designation of HMWB, Chap. 9 on the definition of Maximum Ecological Potential and Chap. 10 on the definition of Good Ecological Potential. Finally, Chap. 11 provides concluding remarks from the synthesis of the thirty-four HMWB case studies and an outlook in view of future research and policy developments relevant to the Water Framework Directive and the issue of Heavily Modified and Artificial Water Bodies.

Annex I includes lists of the contributing authors (HMWB WG members and authors of the case study reports). Annex II includes the case study evaluation tables that were used to extract and collate the results and conclusions of the case studies as well as tables on several case study methods.

Tables, boxes and figures originate either from the HMWB case study reports (as references to sources indicate) or they have been compiled from the case studies by the authors of this book. Reference to the different case studies is made according to the name of the case study and country of origin and not according to the authors of the respective case studies. This has been considered appropriate in the context of this document in order to show the differences between national approaches and methodologies.

[15] WFD CIS Guidance Document No. 2 (2002) Identification of Water Bodies. December 2002.
WFD CIS Guidance Document No. 3 (2002) Analysis of Pressures and Impacts. December 2002.
WFD CIS Guidance Document No. 10 (2003) Rivers and Lakes – Typology, Reference Conditions and Classification Systems. March 2003.

Chapter 2

Guidance on Heavily Modified and Artificial Water Bodies

This chapter describes the identification and designation process for HMWB and AWB as agreed in the CIS guidance document on HMWB and AWB (WFD CIS Guidance Document No. 4 2003). Figure 2.1 illustrates the overall stepwise approach as developed and agreed by the HMWB Working Group followed by an explanation of the individual steps.

We would like to draw the attention of the reader to the fact that the HWMB case studies actually tested an earlier draft version of this process as proposed in the support material (guidance papers) of the HMWB WG for conducting the case studies. This early draft version of the process is presented in Fig. 2.2, and the main differences to the final agreed process of Fig. 2.1 are pointed out in the end of this chapter. Through the process of drafting and finalising the CIS guidance document on HMWB and AWB, the process of identification and designation was significantly revised on the basis of the experience from the HMWB case studies and the discussions within the HMWB Working Group.

According to the final CIS guidance document on HMWB and AWB, the following steps should be taken in the HMWB and AWB identification and designation process (WFD CIS Guidance Document No. 4 2003):

- Step 1: Distinct water bodies are to be identified and described according to the EC horizontal guidance on water body identification.[1] Water body identification is an iterative procedure with possible adaptations in later stages of the designation process (mainly after the provisional identification of HMWB, Step 6). The water body identification has to be done for all surface waters (natural, HMWB and AWB) and is significant because water bodies are the units for which status is being assessed, objectives established and achievement of objectives of the WFD checked.
- Step 2: The WFD gives distinct definitions for AWB and HMWB (Art. 2(8) and Art. 2(9), respectively). In this second step, it should be identified whether the water body concerned has been "created by human activity". If this is the case, EU Member States will have the option to identify it as AWB and consider it for designation or in some circumstances, identify it as a natural water body. Where the intention is to designate as AWB, the first designation test (Step 7) is not relevant and AWB should continue directly with the second designation test (Step 8).

[1] WFD CIS Guidance Document No. 2 (December 2002).

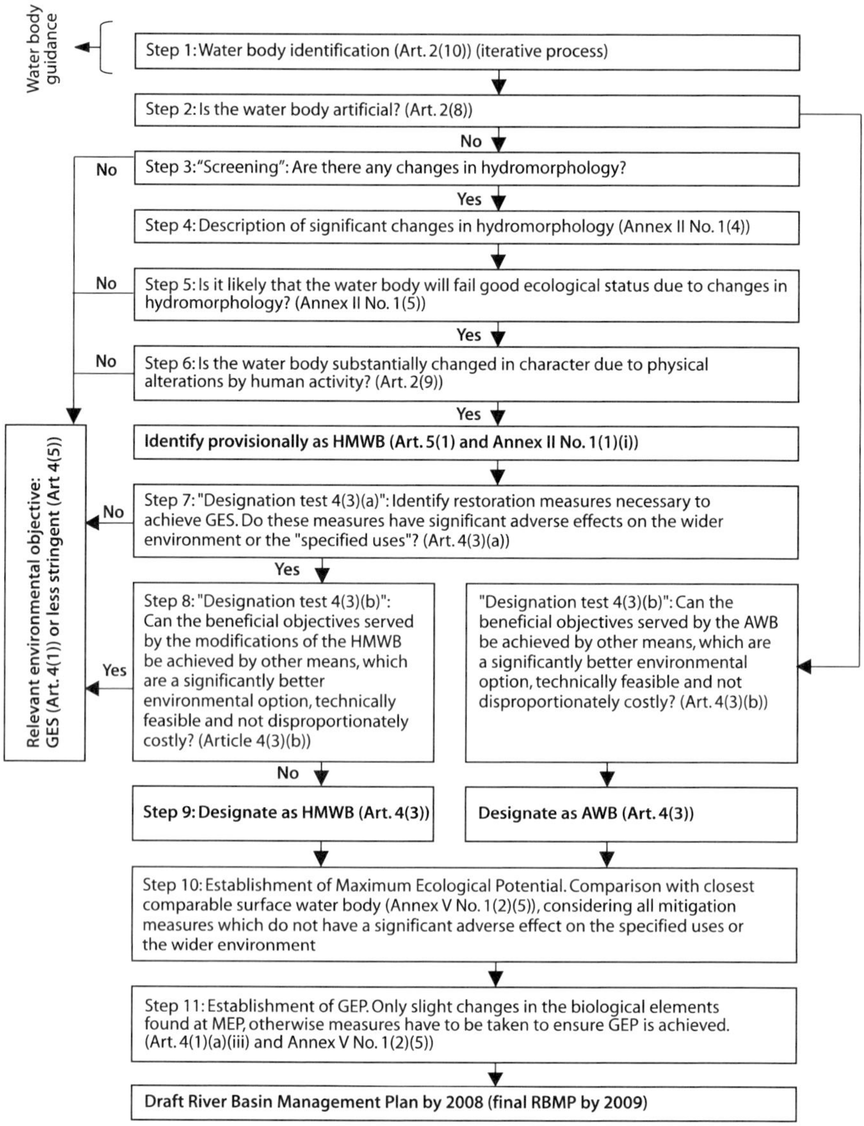

Fig. 2.1. HMWB and AWB identification and designation process (January 2003). *Source:* WFD CIS Guidance Document No. 4, 2003 (final agreed version of January 2003)

- Step 3: A screening process is proposed to reduce effort and time in identifying water bodies that should not be considered for the HMWB designation tests. This will include those water bodies that are likely to fail to achieve GES but show no hydromorphological changes. This step is part of the WFD Annex II (1.4) assessment of pressures.

- Step 4: For those water bodies that have not been "screened out" in Step 3, significant changes in hydromorphology and resulting impacts should be further investigated and described. This includes the description of hydromorphological changes and the assessment of resulting impacts. This step is part of the WFD Annex II (1.4 and 1.5) assessment of pressures and impacts.
- Step 5: Based on the information gathered in Step 4 and an assessment of the ecological status of the water body, the likelihood of failing to achieve Good Ecological Status (GES) (or an estimate of what GES may be, based on current knowledge) should be assessed. Within this step, it has to be assessed whether the reasons for failing the GES are hydromorphological changes and not other pressures such as toxic substances or other quality problems. This step is part of the WFD Annex II (1.5) assessment of impacts process to be completed by 22 December 2004.
- Step 6: The purpose of this step is to select those water bodies where the changes in hydromorphology result in the water body being *substantially* changed in character. Such water bodies can be provisionally identified as HMWB. The remaining water bodies likely to fail GES, which are not *substantially* changed in character, will be identified as natural water bodies. Environmental objectives for such water bodies will be GES or other less stringent environmental objectives.
- Steps 7-8-9: Where Member States wish to designate a water body as heavily modified they must consider them for the designation tests specified under Article 4(3)(a) and Article 4(3)(b). Artificial water bodies are only considered for the test under Article 4(3)(b). In the first "designation test" (Step 7) necessary hydromorphological changes ("restoration measures") to achieve "Good Ecological Status" should be identified. In the first test it has to be assessed whether these "measures" have significant adverse effects on either the "specified uses" or the "wider environment". If they do, then the second designation test (Step 8) is to be carried out.

  The second designation test consists of several sub-tests. Firstly, "other means" to achieve the beneficial objective (e.g. replacement of surface water for drinking water supply with groundwater) are to be considered. Then, it has to be assessed whether the "other means" are (*a*) technically feasible, (*b*) a better environmental option and (*c*) not disproportionately costly. If any of the sub-tests *(a)*, *(b)* or *(c)* are negative, the water bodies may be designated as heavily modified (Step 9). If either the mitigation measures have no significant adverse effects (see Step 7) or if "other means" can be found that fulfil the criteria *(a)*, *(b)* or *(c)* (see Step 8), the water body must *not* be designated as heavily modified and the relevant environmental objective would be GES or a less stringent objective.
- Steps 10-11: These steps are not part of the designation process. However, they are relevant to AWB and HMWB only. They concern the definition of reference conditions and the setting of the environmental quality objectives for HMWB and AWB. In Step 10, the reference condition for HMWB and AWB, the Maximum Ecological Potential (MEP), is defined. Based on the MEP, the environmental quality objective, the Good Ecological Potential (GEP), is defined (Step 11).

The information gathered in the different steps (1–11) summarised above contributes to the River Basin Management Plan (RBMP). The RBMP will contain programmes of measures (WFD Art. 11) that are required to ensure the achievement of the environmental objectives for natural, HMWB and AWB. According to the CIS guidance document on HMWB and AWB, in following the process of Fig. 2.1, it is clearly important to avoid unnecessary and superfluous administrative actions. For example, it will not always be necessary to undertake the assessment for each individual water body. Indeed in many situations, it may be more effective to apply the tests to a group of water bodies where the environmental concerns and specified uses are similar. For example, for a river modified for navigation it may not be helpful to apply the process to individual water bodies. A larger scale assessment may produce a more effective and more complete assessment (WFD CIS Guidance Document No. 4 2003).

There are some distinct differences between the final agreed process on the HMWB and AWB identification and designation of Fig. 2.1 and the earlier draft process of Fig. 2.2:

First, the final process of identification and designation, as included in the CIS guidance document on HMWB and AWB, entails an additional step for screening out early in the process those water bodies that may fail to achieve GES but show no hydromorphological changes (Step 3 of Fig. 2.1).

Secondly, the steps leading to the provisional identification of HMWB are more explicitly described than in the draft process of Fig. 2.2.

Thirdly, the final process of Fig. 2.1 provides a revised timetable for the identification and designation of HMWB and AWB as compared to the draft process of Fig. 2.2. Actually, it became obvious from the results of the HMWB case studies that the timetable of Fig. 2.2 would clearly not match the requirements for the RBMPs when it comes to assessing measures and their costs. In the drafting process of the CIS guidance document on HMWB and AWB, a revised timetable was therefore considered essential to achieving stronger integration of the HMWB and AWB process with the (first cycle of) production of the RBMPs until 2009. Provisional identification of HMWB and AWB should be complete by December 2004 (WFD CIS Guidance Document No. 4 2003). As a general rule for provisionally identified HMWB, Steps 7–11 (of Fig. 2.1) and the assessment of the risk of failing the GEP objective should occur as soon as possible before December 2008 (WFD CIS Guidance Document No. 4 2003). If a designated HMWB or AWB does not meet the GEP objective, then a Programme of Measures or a case for derogation has to be developed by December 2008. This allows one year for consultation of the draft RBMP before publication of the final plan in 2009 (WFD CIS Guidance Document No. 4 2003).

Finally, in the final process of identification and designation of Fig. 2.1, specific steps have been included regarding the operation of the process for AWB (Sect. 5.8 of the CIS final guidance on HMWB and AWB deals explicitly with this issue, WFD CIS Guidance Document No. 4 2003). It is clear from the text of the Directive that the designation tests of Article 4(3) apply to AWB as well as to HMWB. However, the CIS guidance document recognised that the interpretation of Article 4(3)(a) in relation to AWB is problematic. In order to undertake the Article 4(3)(a) designation test, the restoration measures necessary to deliver GES must be identified. This is not possible for AWB because they were created in a location where no significant water existed before, and therefore the High Ecological Status (HES) natural condition would be "dry land" and a sensible GES could not be derived. Consequently, it should be assumed that test 4(3)(a) does not apply to AWB. The second "designation test 4(3)(b)" does not impose interpretation difficulties when applied to most AWB and should be used as a designation test. Consequently, when designating AWB, it should be considered whether

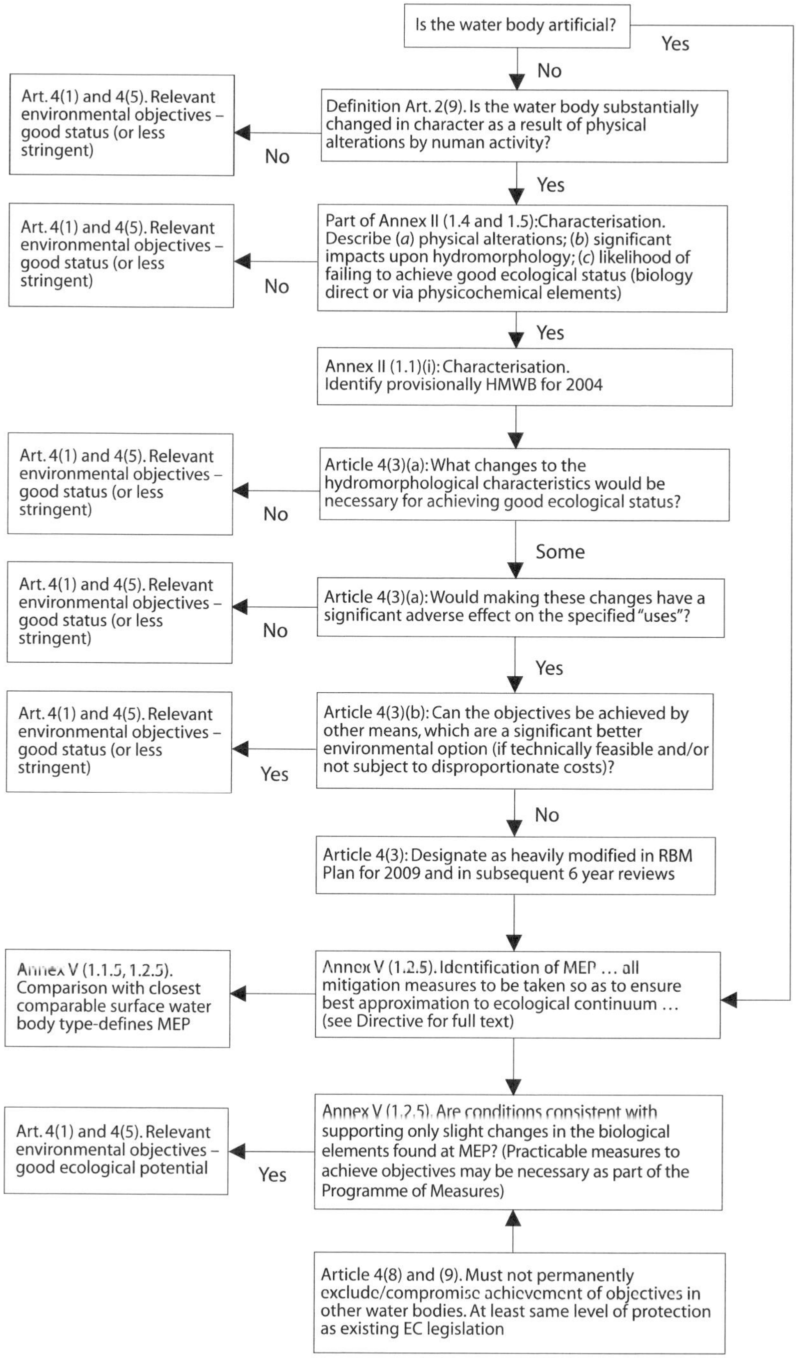

Fig. 2.2. Draft HMWB and AWB identification and designation process (October 2000). *Source:* Support material (guidance papers) agreed by the HMWB WG for conducting the case studies (October 2000)

there are "other means" which can deliver the beneficial objectives of the AWB (WFD CIS Guidance Document No. 4 2003).

The final identification and designation process of Fig. 2.1 was agreed on the basis of knowledge acquired through the activities of the HMWB CIS Working Group until the end of 2002. It should be pointed out that in the future the process displayed in Fig. 2.1 may require adaptation in light of further experiences while implementing the provisions of the Water Framework Directive for HMWB and AWB in the Member States.

Chapter 3

Selection of Case Studies

As already mentioned, an empirical approach using case studies was considered necessary for the CIS Working Group on HMWB and AWB since there was hardly any experience of HMWB and AWB identification and designation in the Member States.

Members of the Working Group (WG) on HMWB were therefore asked to contribute case studies that they considered appropriate and helpful for the further work of the WG in terms of the development of a guidance document on HMWB and AWB. No selection criteria were set for the case studies, and there was no strategic selection of case studies based on criteria such as area, modification type, water body category etc. The main reasons for choosing case studies were: (1) availability of data and funds as well as (2) ongoing projects of relevance for the HMWB process.

In total, thirty-four case studies were carried out. Twenty-eight of them dealt with river systems, only three with lakes, two with transitional waters and one with coastal waters (see Table 14.1 in Annex II). Some of the river case studies included on-stream lakes. Therefore, it is important to point out that the results and conclusions of this document are mainly drawn from experiences with river systems.

Similarly, there is a bias towards medium and large catchment areas.[1] According to Table 14.2 (in Annex II), most of the study areas are medium (11 case studies) or large (12 case studies) catchment areas. Some study areas (7 case studies) are very large and only three are small catchment areas. For the case study on the Baltic coast (S), no catchment size was indicated.

The types of specified uses and physical modifications of water bodies reflected in this book also depend on the selection of the case studies. Certain combinations of uses and water categories might therefore be under-represented (e.g. ports and harbours).

3.1 Overview of the Case Studies

This section gives an overview of the thirty-four HMWB case studies providing basic information on their location (also shown in Fig. 3.1),[2] catchment size,[3] the main pres-

[1] For the size characterisation, the definition of the WFD Annex II (1.2.1) System A for rivers has been used. According to this definition, a very large catchment equals >10 000 km^2, large catchment equals 1 000–10 000 km^2, medium catchment equals 100–1 000 km^2, and small catchment equals <100 km^2.

[2] The location of the HMWB case study areas in Fig. 3.1 is not represented with precision but should be considered as approximate. Most case study areas are more extended than the points on the map.

[3] See Table 14.2 in the Annex for details on catchment size and length of the case study areas.

Fig. 3.1. Map of Europe showing the location of the thirty-four HMWB case study areas. Note: The numbers correspond to the case study numbers given in this chapter

sures affecting the study areas as well as any other information considered important to set the case studies into context. Each case study is described separately in the following paragraphs based on information from the case study reports.

1. *Danube River (A).* The Danube River is one of the largest rivers in Europe (total course 2 850 km). The Austrian case study on the Danube River focuses on the impounded Austrian Danube section between Greifenstein and Vienna (50 km length of river). The case study comprised a combination of different uses; navigation, flood protection, urbanisation and hydropower generation are the main uses. The case study area is characterised by two hydropower stations, bank reinforcement and hydrologically isolated large floodplain areas. Another important aspect of this case study is that a variety of rehabilitation measures have already taken place, resulting in improvements with regard to the ecological status.

2. *Bregenzerach River (A).* Bregenzerach is a medium-sized alpine river and tributary to Lake Constance. This case study served as an example for a river significantly impacted by hydropower peaking and diversion-type power stations, which cause several hydromorphological changes such as reduced flow in the riverbed. A variety of rehabilitation measures (e.g. re-regulating reservoirs and a flow management system introduced to reduce peak amplitudes) have resulted in improvements to the ecological status of the upstream river section, while the downstream part is still considerably affected by hydropower peaking.
3. *Wienfluss (A).* This case study focused on the Wienfluss, which runs through the city of Vienna with a population of about two million inhabitants. The river section of the Wienfluss is about 34 km long with a catchment area of 94 km^2. The focus of the case study concentrated on alterations for flood protection that have an impact on the Wienfluss river system in terms of channelisation and impoundment. Besides flood protection, urbanisation is another main use in the case study river section. The history of river regulation in the Vienna section of the Wienfluss is closely related to urban development, which started in the 19th century on a large scale. Additionally, the Wienfluss was chosen as a case study due to the variety of rehabilitation measures that have already been carried out and have consequentially led to improvements of the ecological conditions within this river section.
4. *Dender River (B).* This case study covered the lowland Dender River as well as its tributaries, Mark and Bellebeek. The Dender has a large catchment, which is part of the catchment area of the Schelde River. The Dender is a border-crossing river between the region of Flanders and Walloon in Belgium; therefore, collaboration between the regions is required. The case study area is influenced by different types of specified uses such as flood protection, navigation, urbanisation and agriculture as well as various physical alterations such as canalisation, straightening and dams. Due to the geology of the case study area, the stream discharge has a quick response to rainfall events. However, exceptional discharges of e.g. 60 $m^3 s^{-1}$ in wintertime or 0 $m^3 s^{-1}$ in summertime are caused by canalisation and the locks.
5. *Dhünn River (D).* The Dhünn River is a medium-sized river located in the mountainous region of north-western Germany. It is the major tributary to the Wupper River, which, shortly after their confluence, enters the Rhine River. The upper 60 km^2 of the Dhünn catchment is collected by a deep reservoir for drinking water supply for the industrial center of the lower Rhine valley. The case study was chosen to assess whether the stretch downstream of the dam has to be designated as heavily modified or whether this designation should be restricted to the dam area.
6. *Mulde River (D).* The Zwickauer Mulde River (167 km) is an important tributary of the Elbe River, rising from the central mountains in middle-east Germany. It comprises a large catchment with different stream types and landscape areas. It flows through a historically industrialised region impacted by various specified uses such as urbanisation, water supply with several reservoirs, hydropower generation as well as former uranium mining. For the purposes of the case study, a study area 65 km long, which included the city of Zwickau, was selected.
7. *Elbe River (D).* The Elbe is one of the largest rivers in central Europe with its springs in the central highlands of the Czech Republic (total length of 1 091 km). Today, the entire German Elbe is a national waterway, while Hamburg, which is located on the Elbe, is the most important harbour in Germany. For the case study, two representa-

tive sections were selected: one on the upper and one on the middle Elbe. The upper Elbe stretch from the Czech-German border up to Pirna (34 km) reflects typical impacts of navigation, while on the middle Elbe stretch between Magdeburg and Geesthacht (250 km) flood protection prevails. Throughout the entire Elbe, water quality remains a problem.

8. *Lahn River (D).* The Lahn River (244 km) is a tributary to the Rhine River and represents a typical low mountain range river. Main uses in this case study are hydropower generation and navigation. Overall, more than twenty-two river hydropower stations and over fifty weirs, many of them not passable for aquatic organisms, are located along the river. Navigation becomes more important within the last 150 km before merging into the Rhine, and therefore this downstream section has been strongly developed (canalisation, river bank fixation, weirs, and sluices). In 1981, traffic of goods was discontinued, and nowadays passenger and recreational navigation prevails.
9. *Seefelder Aach River (D).* The Seefelder Aach in south-western Germany is a 54 km tributary to Lake Constance. The Seefelder Aach was chosen as a case study because of strong impacts due to intense hydropower production. Twelve hydropower plants and five additional weirs are disturbing the migration of aquatic fauna, especially of the lake trout *(Salmo trutta)*. Furthermore, intensive agricultural land use, including fruit and vine cultivation, have already led to large-scale alterations with regard to land drainage and river regulation.
10. *Ruhr River (D).* The Ruhr, a tributary of the Rhine, is situated in the mid-western part of Germany with a length of 219 km. The "Ruhrgebiet" located in the western part of the catchment is one of the largest industrial areas in Europe, including the large inland harbour of Duisburg. The water of the Ruhr serves as a drinking water resource for over five million people and as an industrial water supply. To enable this water supply, fourteen reservoirs with dams have been built in the upper catchment also used for hydropower generation. Hence, this case study focused on the impacts of water supply. For the purposes of the case study, an investigation area was selected at the lower River Ruhr, including three impounded lakes and the cities Dortmund and Bochum.
11. *Lozoya River (E).* The Lozoya River is situated towards the centre of Spain. It rises in the central mountain range (*Sistema Central*) at a high altitude and cuts through steep relief joining the Jarama River at 710 m a.s.l. The Jarama is a major tributary of the Tajo, the longest river in Spain with the greatest storage capacity. Lozoya is an example of a Spanish river that has been significantly altered by the construction of a series of reservoirs, serving nearly 50% of the total water supply for the metropolitan area of Madrid and to a lesser extent for hydropower generation (five hydropower plants). Nowadays, roughly 50% of the river length is taken up by reservoirs. Other specified uses, such as adjacent agriculture, forestry and human settlements, are of lesser significance in relation to the physical impacts on the water body.
12. *Lake Kemijärvi (SF).* Lake Kemijärvi is the largest natural lake within the catchment of the Kemijoki River, the longest river in Finland situated in Lapland. Lake Kemijärvi is the most heavily regulated lake of Finland, and most of the harmful effects of regulation are obviously visible. The main objectives for the regulation are hydropower production and flood protection. Other human activities in the

drainage area include forestry, farming, husbandry and fish farming, while some pollution is also caused by the wood pulp industry and settlements. Water level regulation of Lake Kemijärvi has a maximum amplitude of 7 m, which is the largest regulation amplitude encountered in Finnish lakes, and has caused harmful impacts on the littoral vegetation, the benthic invertebrate fauna biomass and fish stocks. The lake's strong spring floods have been cut off by water level regulation, allowing for higher winter discharges for hydropower. On the other hand, summer water levels have been raised due to higher storage capacity. The case study on Lake Kemijärvi was within the context of a larger study aiming to identify Finnish-regulated lakes based on water level fluctuation and biological characteristics.

As a part of this case study, water body identification was carried out in the main stream of the Kemijoki River downstream of Lake Kemijärvi. This part of the river consists of eight hydropower plants and run-off-river impoundments. The pressures of human activities on the Raudanjoki River, a large tributary of the Kemijoki River below Lake Kemijärvi, were also identified.

13. *Authie River (F).* The Authie River in northern France is 95 km long and flows into the English Channel. It is mainly affected by a series of water mills for different uses (electricity, fish farms) with small dams of 0.5–2 m height which hinder fish migration. Further impacts derive from large-scale agriculture, urbanisation and recreation. The low and central valley of the river has been proposed as the site of the Natura 2000 network (potentially protected area).
14. *Sarre River (F).* From its source in the Vosges mountains (altitude of 700 m) to the German border (altitude of 194 m), the Sarre has a length in excess of 130 km and includes various types of water courses: medium-sized mountain torrents, sectors of free meanders in the valleys and more restricted water courses in the form of fixed meanders. Human uses include silviculture, extensive livestock farming and urbanisation. Along the Sarre, the principal physical alterations are diversion of part of the discharge for the channel supply, modification of the riverbed by transverse weirs (mills dam), modification of banks, and channelisation increased by urbanisation (downstream sections).
15. *Rhone River (F).* The Rhone River valley is the largest hydrosystem of the Rhone-Mediterranean-Corsica basin (covering 70% of the total 130 000 km^2 basin area). Its source is in Switzerland, where its catchment basin is 8 000 km^2. The Rhone was chosen as a case study area because it has a relatively complete range of anthropogenic pressures and uses typically seen in various rivers of southern France. Moreover, it is one of the most studied rivers in France and has a dense monitoring network of water quantity and quality measurement stations. Since ancient days, the Rhone has been a major hub of human activities, some of which are of critical importance at the national level (e.g. hydropower production and inland navigation). Today, its valley is still of major social and economic importance mainly concentrated around the Lyons conurbation and its industry. The physical alterations are linked to hydropower, navigation, urbanisation and flood defence. The Rhone draws the attention of numerous stakeholders directly or indirectly involved in strategic discussion and/or decision making for concerted management of the resource.
16. *Nestos River (GR).* The Nestos River is 234 km long. It is an international river with its source in south Bulgaria, flowing through Greece and finally discharging into the Agaeis at the ecologically important Nestos Delta. The case study focused on

the Greek part of the river (130 km), which has two large hydropower reservoirs also serving water supply and irrigation. A third hydropower reservoir is under construction further downstream. The Nestos Delta is under the protection of the Ramsar treaty for wetlands, but it has been heavily impacted by agriculture, land drainage and river straightening, which has resulted in a discontinuation of recent delta formation.

17. *Haringvliet Estuary (NL).* The case study represents an example of a modified transitional water body. The Rhine and Meuse Rivers form a combined estuary in the south-west of the Netherlands. The northern outlet of the estuary is the Rotterdam Waterway. The southern outlet is the so-called Haringvliet Estuary, which was the focus of the case study. After completion of the Haringvliet Dam in 1970, the area changed from a dynamic brackish tidal inlet into a semi-stagnant freshwater area. The flow regime is regulated by sluices to ensure minimum water flow in the Rotterdam Waterway. Currently, there are plans to open/remove the Haringvliet Dam in order to restore the estuary system. The primary uses in the case study area are navigation, flood protection and agriculture.
18. *Hagmolenbeek-Hegebeek River (NL).* Hagmolen-Hegebeek is a small trans-boundary stream (length 27 km) located in the eastern part of the Netherlands at the German border. The case study was chosen because it represents a typical lowland stream of the Netherlands significantly impacted by agriculture. The stream has been changed by channelisation, the construction of weirs, and drainage in the catchment to optimise the hydrological conditions for agricultural use. Since the construction of the Twente Canal (1936) for navigation, the Hagmolen-Hegebeek has been cut off from the catchment of the stream Regge and now flows directly into the Twente Canal.
19. *Lake Loosdrecht (NL).* Lake Loosdrecht (10 km^2) was selected as a typical example of a Dutch peat lake and is part of a system of shallow interconnected lakes in the centre of the Netherlands. While former industrial peat mining created the preconditions for the creation of the lake, natural processes formed the present lake. It is heavily impacted by recreation, especially recreational shipping and the construction of unnatural embankments to prevent further erosion of the peat banks. A further problem is eutrophication and water quality.
20. *Lake Veluwerandmeren (NL).* The Veluwerandmeren is a Dutch shallow freshwater lake system with a medium catchment influenced strongly by hydromorphological changes. It consists of four connected lakes: Veluwemeer, Drontenmeer, Wolderwijd, and Nuldernauw. These were created by the reclamation of a polder and the construction of dikes at a former estuary. Due to many hydromorphological impacts, Veluwerandmeren has no stable period of relatively undisturbed ecological functioning. The main use of the lake is fisheries and recreation accompanied by heavy recreational shipping, as well as water supply (for irrigation and industrial processes). It has been designated as a protected area according to the EU Birds Directive.
21. *Suldalslågen River (NO).* The Suldaslågen River is located in south-western Norway and represents a typical example of a river affected by hydropower generation in Norway. Within the catchment there are two large hydropower schemes, each with several reservoirs and power stations. Another typical feature is the transfer of water across catchment borders, which is diverted from other rivers to the Suldaslågen, increasing the catchment area from 1 460 km^2 to 2 140 km^2. Since the Suldaslågen is

one of the most famous Atlantic salmon rivers in Europe, the impact on salmon fisheries is of particular concern. The case study focused on the lower part of the river, a stretch of 22 km from Lake Suldalsvatn to the river mouth.

22. *Beiarelva River (NO).* The Beiarelva River (50 km) is situated in northern Norway at the margin of the second largest plateau-glacier (Svartisen) in a high mountainous area. The Beiarelva River is affected by the diversion of water from its upper tributaries, resulting in a reduction of discharge volume and of the catchment area from 859 km^2 to 766 km^2. The impacts result from hydropower production in an adjacent river. Serious problems affecting the Atlantic salmon population arose from the infestation of the river by salmon parasites.
23. *Umealven River (S).* The Umealven (450 km) in northern Sweden was chosen as a case study because it is representative of the rivers that drain the Scandinavian mountain range and empty into the Gulf of Bothnia. Umealven is heavily affected by hydropower but comparatively unaffected by other human activities and therefore a good example to study the effects of hydropower. The hydropower stations (run-off-river impoundments) use almost the entire fall height of the river, from the storage reservoir at 520 m a.s.l. to the sea. Reference conditions were derived from the neighbouring Vindel River, which is of similar size and has similar hydrology but is still free flowing.
24. *Emån River (S).* Emån is the largest river in south-eastern Sweden and is heavily affected by over forty hydropower plants along its course. The study focused mainly on the sub-catchment area of the short Solgenån River (23 km) in the northern part of the Emån catchment. In its basin, there are three large hydropower plants hindering the migration of salmon. Other anthropogenic impacts are relatively insignificant. Land use is mainly restricted to pasture, and the surrounding vegetation consists predominantly of wetlands and coniferous forest in the upper stream sections.
25. *Dalälven River (S).* The Dalälven case study in Sweden focused on the fish fauna and the specific problems caused by regulation and hydropower for fish, including the release of salmonid fish for compensation purposes. The river is highly affected by modifications due to approximately five hundred dams and thirty-nine hydropower plants. Regulation dams and constructions for hydroelectric power affect lakes and rivers in the drainage basin as well as the smaller streams. Ditching forests and mires, road culverts, and log-driving channels and dams have profound effects on smaller streams in the catchment.
26. *Baltic Sea Coast (S).* This case study was selected as an example for Heavily Modified Water Bodies in coastal areas up to 1 nautical mile off the main coast. The Stockholm Archipelago is one of the most complex parts of the Baltic Shore with over 24 000 islands, rocky islets and skerries. Small-scale variations in landscape and ecosystems according to exposure, salinity and currents characterise the study area. The study area comprised the southern part of Stockholm Archipelago. This has areas which range from heavily modified to less modified. The primary uses specified for this study area are urbanisation, navigation and recreation, leading to physical modifications such as straightening of the shoreline, erosion protection, dredging and infrastructure along the shoreline.
27. *Sankey Brook (UK).* Sankey Brook (127 km) is a tributary to the Mersey River located east of Liverpool, England. It is representative of many rivers located in the

industrial regions of the UK. Sankey Brook is subject to a mixture of specified uses such as urbanisation, industry, flood defence and agricultural development. Impacts include land drainage, poor water quality and general degradation of instream and riparian habitats. Many physical modifications stem from historical industrial development, especially mining.

28. *Great Ouse River (UK).* The Great Ouse catchment covers much of East Anglia (Eastern England) and represents a large-sized, heavily regulated lowland river. Much of it has been heavily engineered for flood defence and land drainage as well as for navigation purposes. Modifications include completely artificial cut-off channels, channel re-alignment and resectioning, bank reinforcement, weirs/locks, and loss of floodplain channel diversity. As a result of drainage, fens were transformed from wetland with raised islands of clay into some of the most productive arable land in the UK. Overall, despite the extensive human influence on the landscape, parts of the area have been designated as Special Protection Areas (SPA), Sites of Special Scientific Interest (SSSIs) and Ramsar sites.
29. *Tame River (UK).* The Tame River basin in central England is an example of a medium catchment with widely varying land use, river use, river modification and ecology. The main Tame River runs through the heavily urbanised area of Birmingham and represents an example of a degraded urban river with heavy urbanisation at the top of the catchment. Urbanisation has led to higher flood peaks as well as pollution, which combined with poor habitats, has led to poor ecological conditions. Some of the River Tame's tributaries are less modified and are good examples of degraded rivers with potential for rehabilitation.
30. *Kennet River (UK).* The Kennet River is the largest tributary (70 km) of the Thames River in south-central England. It represents a river that has many uses and is exposed to many pressures from human activity. These include navigation, flood defence, groundwater abstraction, land drainage, and management for agriculture, fisheries (commercial and recreational), heritage such as old mill structures, and some degree of urbanisation. It also receives effluent from sewage treatment works, and its catchment, especially the floodplain area, is high quality agricultural land. Despite these pressures, the ecological status of the river is generally considered to be good. The Kennet valley also contains nationally important wildlife sites and habitat types; there are fifty-five Sites of Special Scientific Interest (SSSIs) designated by English Nature within the Kennet catchment.
31. *Forth Estuary (UK).* The Forth Estuary represents one of the two case studies for transitional waters. The Estuary and the Firth River of Forth together form a key segment of the east coast of Scotland, stretching from the centre of the country eastwards past Edinburgh to join the North Sea. Characteristic of this classic trumpet-shaped estuary with a length of about 45 km are high tidal ranges (up to 5 m) and large areas of intertidal mud. The margins of the Forth Estuary are densely settled with four major urban settlements. Hence, the main impacts are linked to urbanisation, flood defence and loss of intertidal zones.
32. *Dee River (UK).* The Dee River (Galloway) has a large catchment and is situated in the southern uplands of Scotland. Typical for this area is a wide network of streams and natural lakes. The area has a very low population density with about 0.23 persons per hectare with pasture and forestry as main land-use patterns. The principal specified use in the Dee River is hydropower. The Galloway hydropower scheme

comprises a sequence of six hydropower stations with a total installed capacity of 109 MW.

33. *Tummel River (UK).* The catchment of the Tummel River is located in the northern highlands of Scotland. It covers an area of 1 713 km^2 and reaches a peak altitude of 1 083 m a.s.l. The Tummel study area has a very low population density (less than 0.10 persons per hectare). The single use specified in the Tummel River basin is large-scale hydropower generation by a scheme of five hydropower stations and an extensive channel system to direct water from other catchments towards the stations. Habitat creation associated with some water bodies in the hydropower scheme area is of acknowledged nature conservation value. These parts of the water bodies have been designated as Sites of Special Scientific Interest (SSSIs).
34. *Lagan River (UK).* The Lagan has a medium catchment, which is the most densely populated area in Northern Ireland including the city of Belfast. It drains some of the most productive agricultural lands in Northern Ireland. The estuary of Lagan (Belfast Lough) is characterised by the Belfast harbour and is hence heavily changed and subject to dredging activities. The main use in the Lagan catchment is urbanisation in the lower catchment while the main physical modifications result from industry, flood defence and navigation.

Chapter 4

Identification of Water Bodies

4.1 Introduction

As shown in Fig. 2.1, the first step that should be taken in the HMWB and AWB identification and designation process is the identification and description of distinct water bodies.

According to the Water Framework Directive, Art. 2(10), a body of surface water means a discrete and significant element of surface water such as a lake, a reservoir, a stream, river or canal, part of a stream, river or canal, transitional water or a stretch of coastal water. All waters must be assigned to a water body. A river basin is subdivided into several water bodies, and one or several river basins can form a river basin district.

As stated in Annex II of the WFD, the (identified) surface water bodies within a river basin district shall be identified as falling within either one of the natural surface water categories (rivers, lakes, transitional, or coastal waters) or as Artificial Water Bodies or Heavily Modified Water Bodies. For each surface water category, the relevant surface water bodies within a river basin district shall be differentiated according to types using either typology system A or typology system B of the WFD Annex II 1.2.

Water bodies form the management and classification units of the Water Framework Directive. Consequently, it is important that the identification of water bodies is carried out at an appropriate scale to allow for an ecological classification that is detailed enough to reflect the current status. Once the location and the boundaries of a water body have been identified, environmental objectives are to be set that apply to the entire water body. Thus, identification of water bodies is a crucial step with far-reaching implications for the implementation of the WFD. Therefore, a clear mechanism and method for the identification of water bodies is necessary and should receive special attention.

The final guidance document on HMWB and AWB (WFD CIS Guidance Document No. 4 2003) does not include specific guidance on the identification of water bodies but refers to the EC horizontal guidance document on this issue. The EC horizontal guidance document on the identification of water bodies (WFD CIS Guidance Document No. 2 2002) has been produced as a part of the WFD Common Implementation Strategy and should be used for the identification of water bodies. The horizontal document was finalised after the completion of the HMWB case studies; therefore, its final version could not be taken into consideration in the process of identifying water bodies within the case studies. Nonetheless, it should be pointed out that the experience gained in the HMWB case studies on water body identification partly contributed to the development of the identification criteria relevant to HMWB and AWB proposed in the EC horizontal guidance document.

4.2 Criteria for the Identification of Water Bodies

4.2.1 Overall Application of Identification Criteria

The support material (guidance papers) agreed by the HMWB WG for conducting the case studies proposed four criteria for the identification of different water bodies: WFD typology system A and B, hydromorphological units, type-specific biological conditions and effective management units. Therefore, the identification of water bodies in the HMWB case studies was mainly based on these criteria. In the UK, existing water quality stretches (based on effective water quality management units) were also used as criteria. The identification criteria finally agreed within the Common Implementation Strategy are illustrated in the EC horizontal guidance document on the identification of water bodies (WFD CIS Guidance Document No. 2 2002).

WFD Typology Systems A and B

The basis of defining ecological status is the identification of water body types using a system of typology as defined in Annex II of the WFD. The Directive defines two systems of typology for rivers, lakes, transitional and coastal waters: system A and system B. A surface water body must not cross the boundaries between surface water body types. It must be of one type or another since one purpose of characterising surface water bodies is to differentiate them into types (WFD CIS Guidance Document No. 2 2002).

Typology System A

Typology system A makes use of the ecoregions shown in Annex XI of the WFD, altitude typology, size typology and geology as typology descriptors. Only the case study on the Umealven River (S) followed the WFD typology system A for a first identification of water bodies. The Umealven River crosses two ecoregions (borealic upland region/fennoscandian shield ecoregion), and therefore two main groups of water bodies were derived according to the ecoregions. In many cases, especially with small and lowland rivers or lakes (e.g. Dender (B), Seefelder Aach (D), Hagmolenbeek (NL)), system A did not allow for the identification of separate water bodies. In several case studies (e.g. Bregenzerach (A), Lahn, (D), Ruhr (D), Lozoya (E)) differences occurred in one or two factors of system A (altitude, ecoregion) but were not taken into consideration for the identification of water bodies.

In general, system A was characterised in some case studies as too coarse (e.g. a more detailed description of the local geology was often used instead of the three broad geology types given in system A) and as not reflecting adequately the pressures related to the identification of HMWB.

Nevertheless, many case studies (26 out of 34) used the parameters of typology system A to describe their case study area rather than separate it into water bodies.

Typology System B

Typology system B offers an alternative characterisation for water body types according to obligatory factors (altitude, latitude, longitude, geology and size) and optional physical factors (e.g. water depth and substratum composition). Only four case studies (Dender (B), Ruhr (D), Baltic coast (S), and Forth Estuary (UK)) used system B explicitly. In the case of Forth Estuary, the use of the optional factors depth, salinity, substrate, and turbidity led to the identification of three distinct water bodies.

The case studies rarely used system B exactly as set out in the WFD. Instead, many case studies considered a variety of physical factors such as local geology, substrate, depth, slope, energy of flow, salinity and geomorphology, some of which are included in the optional factors of system B. In the Swedish case study on the Emån River, it was pointed out that the use of system B leads to a detailed description of the study area and creates a higher possibility of identifying the differences between water bodies than system A. The authors of the Swedish case study recommended the development of a general method including the obligatory factors and some of the optional factors from system B to make comparisons between physically modified rivers. The optional factors they chose were energy of flow, mean water depth, mean water width, mean water slope, form and shape of main riverbed, river discharge category, transport of solids, and mean substratum composition. The reason cited for choosing these factors is that in the case of modified rivers, the extent of the morphological changes is indicative of the degree of impact on the ecological status.

After the case study evaluation, it remained unclear whether the WFD typology systems should be used for identifying different water bodies. It seems rather the case that water bodies can be differentiated by other criteria and then described using the WFD typology systems. This approach was followed in the case studies on the Nestos River (GR), the Beiarn River (NO) and the Emån River (S). In the case study on the Nestos (GR), water bodies (natural river, regulated river, reservoirs, and delta) were identified according to specified uses and physical alterations and then described according to the WFD typology system A.

Hydromorphological Units

Hydromorphological units take account of major physical features, e.g. separate tributaries, main river stretches between large tributaries or delta formation. In twenty-one out of thirty-four case studies, hydromorphological units (mostly tributaries) were considered for the identification of different water bodies. In the Dender River (B), the merging of the river with important side streams and the winding character of the river served as criteria for water body identification. In the case study on Lake Kemijärvi (SF) (see Fig. 4.1, p. 33), lakes that were isolated from the main lake (and were only connected to it during spring flood before regulation) were regarded as separate water bodies. In the UK case studies on the rivers Dee and Tummel, primary units such as main river stretches, tributaries and standing waters also served as criteria for water body identification.

Type-Specific Biological Conditions

The criterion "type-specific biological conditions" takes account of the degree to which the water body forms a consistent ecological type in terms of ecosystem structure and function. This criterion is based on the requirement of the WFD that it should be possible to define type-specific reference conditions. This criterion was used in only ten case studies for the identification of water bodies, because there was often a lack of biological data or type-specific reference conditions. Nevertheless, it seems to be a valuable criterion especially for lakes/reservoirs of trophic status (e.g. Veluwerandmeren, Loosdrecht (NL)) and transitional waters (e.g. tidal freshwater marsh in the Haringvliet Estuary (NL)).

In the case study on the Dalälven River (S), it was suggested that the limits of water bodies should be ecologically relevant; in this case study with a focus on fish, it was stated that a water body should be large enough to maintain a sustainable fish population. A minimum viable population of, for example, brown trout can be expected to consist of very roughly 1 200 individuals, whereof 500 are of reproductive age. The size of the identified water body would then depend on the population density.

The case study on the Emån River (S) considered the results of the river habitat survey, while the case studies on the Lozoya River (E) and the Lagan River (UK) considered riparian vegetation for the identification of water bodies. In the Kemijöki River (part of the case study on Lake Kemijärvi (SF)), the existence of similar habitats and the possibility of defining MEP/GEP were taken into account for water body identification.

Effective Management Units

The criterion "effective management units" takes account of the socio-economic uses of water, the induced pressures and resulting impacts. Effective management units appeared to be the most important criterion for identifying different water bodies; the criterion was applied in thirty-one case studies. Separation of water bodies turned out to be more straightforward with distinct physical alterations (weirs, dams) than with land use impacts, water abstraction or navigation. Several studies suggested that each separately identified water body should be influenced by only one kind of use (e.g. Bregenzerach (A), Emån (S)).

Some case studies (e.g. Lahn River (D) and Hagmolenbeek River (NL)) also considered administrative borders for the separation of water bodies. In the Dutch case study on the Harmolenbeek River, this decision was based on different water policies and therefore management considerations affecting different water stretches. In the German case studies, however, it was concluded that the subdivision of water bodies according to administrative entities is not practicable or useful.

Other Criteria

Water quality stretches: The UK-EW case studies on the Great Ouse and the Sankey Brook followed an approach based on current practice and available information for the purpose of the case studies. The approach identified separate water bodies according to an existing network of water quality stretches used for water quality classification. Since these water quality stretches are of very detailed scale (approximately every 4–6 km river length), upscaling regarding the appropriateness for HMWB designation was necessary. This was done on the basis of hydromorphological units and effective management units. The case studies did identify that this may not be appropriate in the final implementation of the Water Framework Directive and suggested other approaches that could be followed. A more WFD relevant approach should be developed; this should be based on physicochemical and hydromorphological properties of rivers, over which pressure information would be layed to derive effective management units.

Aggregation and type of data: The case study on the Dender River (B) considered the level of aggregation and type of available data as an additional criterion for water body identification.

4.2.2
Identification Criteria for Different Water Categories

Rivers

Approaches and results for identifying water bodies in rivers differed widely from case study to case study. Two main approaches can be summarised:

- Identification of water bodies primarily based on hydromorphological and physical characteristics, which usually resulted in distinct upstream/middlestream/downstream water bodies as well as tributaries as separate water bodies;
- Identification of water bodies mainly based on effective management units. The consideration of uses, impacts and infrastructure usually resulted in a differentiation of natural/unnatural or regulated/unregulated river stretches and reservoirs.

In case studies on navigation, agriculture or mixed uses, water body identification mainly resulted from the consideration of hydromorphological and physical characteristics. In five case studies (Mulde (D), Haringvliet Estuary (NL), Tummel, Dee and Lagan Rivers (UK)), effective management units were considered as a second criterion for water bodies primarily identified by hydromorphological/physical criteria.

In most case studies on hydropower or water supply, the identification of water bodies was based on effective management units. Five case studies also considered hydromorphological units. Water bodies were identified according to physical alterations into unregulated (up)stream stretches, reservoir/impounded waters and regulated river stretches. The presence of physical alterations (e.g. weir, dam, etc.) usually served to define the boundaries of the identified water bodies. Individual reservoirs were either identified as separate water bodies (e.g. Emån (S)) or the whole system of reservoirs was defined as one water body (e.g. Nestos (GR)).

The identification criterion of type-specific biological conditions was not used in most case studies on rivers. This may be partly explained by the lack of data and difficulties in establishing type-specific biological conditions within heavily impacted rivers.

The case studies also revealed that further clarification is needed on how to identify water bodies within a sequence of hydropower stations/reservoirs, especially in case of run-off-river impoundments. The question as to whether run-off-river impoundments should represent separate water bodies was addressed in the case study on Lake Kemijärvi (SF) and is presented as a comprehensive example in Box 4.1. In the latter, it was concluded that run-off-river impoundments should be considered as one water body mainly because:

- All impoundments belong to the same river type;
- It would be unreasonable to define MEP and GEP for each impoundment separately; and
- A holistic approach would be more practicable since power plants are operationally used as one unit.

Similar conclusions were drawn in the case study of Seefelder Aach (D), where a sequence of seventeen run-off-river impoundments was identified as one water body (WB). In the Swedish case study on the Umealven River with five reservoirs and sixteen run-off-river impoundments, each impoundment was, however, identified as a single water body which resulted in a large number of water bodies. In this case, a process-based approach was used where each water body should be controlled by the same water-level regime (common physical forces).

Lakes

Only three case studies (Lake Kemijärvi (SF), Veluwerandmeren (NL), Loosdrecht (NL)) focused on lake systems. Some of the river studies did consider on-stream lakes (e.g. the case studies on the rivers Tummel and Dee (UK), which focused on a network of lochs and rivers), but these lake systems are not covered in this document.

Box 4.1. Identification of Run-Off-River Impoundments in the Kemijöki River, Finland

Kemijöki River has a length of about 220 km and it has eight hydropower plants and seven run-off-river impoundments.

The middle part of the river (70 km) is not modified by constructions and resembles the natural Kemijöki River. There were three main alternatives for the separation of the main stream of the Kemijöki River into water bodies:

a One Heavily Modified Water Body
b Two water bodies: heavily modified and natural (non-constructed part)
c Seven water bodies: each run-off-river impoundment is one water body and non-constructed part is one water body

The following questions were raised during the water body identification process:

- *Are run-off-river impoundments separate water bodies?*
  - YES: Regulation pattern and problems are different.
  - NO: They belong to the same river type.
  - NO: It is not reasonable to define Maximum (MEP) and Good Ecological Potential (GEP) separately for each impoundment, which is needed in the case of separate water bodies.
  - NO: Holistic approach is more reasonable because power plants are used operationally as one unit and finding cost-effective mitigation measures may require a wider point of view.

  ⇒ CONCLUSION: NO

- *Is a non-constructed part of a river a natural water body?*
  - YES: Some habitats (e.g. shore meadows, erosion banks) typical for the natural Kemijöki River still exist.
  - NO: According to flow analysis, short-time regulation has changed the hydrological regime significantly. The remaining natural part represents only a small fraction of the main channel.
  - YES/NO: Changes in fish fauna are due to the power plants.
  - Only one fish ladder exists in the Kemijöki River (at the lowest power plant)
  - The possibility of applying "less stringent environmental objectives" needs to be considered.

  ⇒ CONCLUSION: NO or more information is needed on the current status of the river.

Source: Case study on Lake Kemijärvi, SF.

The identification of water bodies was straightforward in all lake case studies with one lake being identified as one water body as assumed by the WFD Annex II.

Lake Kemijärvi (see Fig. 4.1) represents the most complex system, which can be divided into a main lake area with maximum water level regulation and several smaller sub-basins with less regulation. Despite these differences in the regulation regime, Lake Kemijärvi was identified as one water body considered as a reasonable management unit. Smaller surrounding isolated lakes were identified each as a single water body

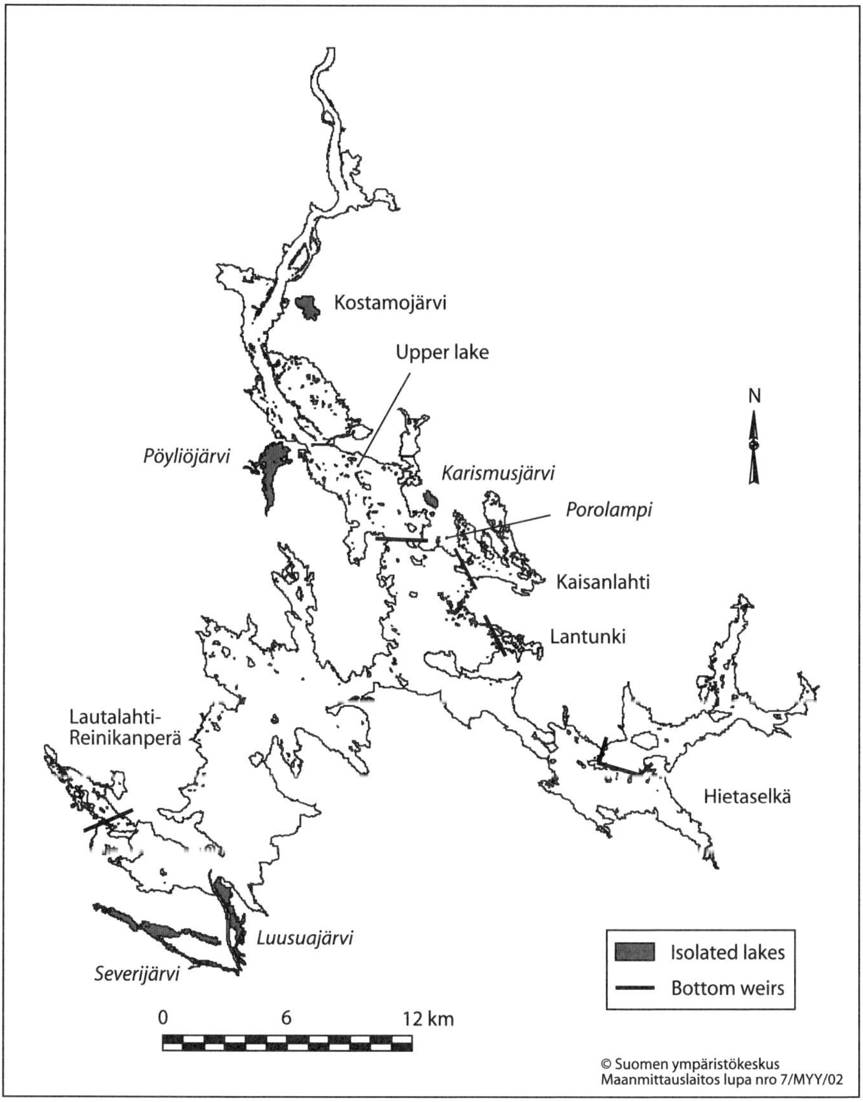

Fig. 4.1. Lake Kemijärvi and isolated lakes, Finland. *Source:* Case study on Lake Kemijärvi, SF

according to hydromorphological units. The Veluwerandmeren (NL), which is a series of connected lakes, was also identified as one water body.

Coastal Waters

The only case study that focused on coastal waters (Baltic coast (S)) did not identify separate water bodies. The water body identification criteria of the support material (guidance papers) of the HMWB WG for conducting the case studies were not considered as appropriate for coastal areas. The main problem, besides the lack of natural borders, was insufficient data on hydromorphological parameters and physical disturbance in the shallow waters of the Baltic Sea. The case study used the existing grid of coastal sea areas and basins (Swedish Meteorological and Hydrological Institute), but this proved to be unsuitable in respect to the scale and impacts of uses. Moreover, according to the authors of this case study, this grid needs to be adjusted to the coastal water boundary of 1 nautical mile as required by the WFD.

Transitional Waters

Two case studies (Haringvliet Estuary (NL) and Forth Estuary (UK)) investigated transitional waters. Each identified three water bodies mainly according to physical factors (e.g. salinity, mixing regime and tidal range in the case of the Haringvliet Estuary (NL)), ecological units (tidal freshwater marsh, intertidal mudflats) as well as according to pressures and impacts (the latter especially in the case of the Forth Estuary (UK)). In the Forth Estuary, the main criteria for water body identification were differences in the types and intensities of human pressures and impacts, backed by differences in morphology and physical conditions such as salinity.

4.3 Results

4.3.1 Overview of Water Body Identification Results in the HMWB Case Studies

The main results on water body identification are summarised in Table 14.3 (see Annex II). Rivers were usually divided into large (e.g. approximately 3–4) water bodies. For instance, in the case of the Emån River (S), a study area of 23 km was divided into four water bodies each 4–6 km long. Some rivers were broken up into smaller water bodies with a maximum of fifty-seven water bodies in the extensive Tummel River network in Scotland (UK). Regarding rivers with large and very large catchments, the results are not necessarily representative, since in the HMWB case studies that focused on rivers with large/very large catchments (e.g. Danube (A), Great Ouse (UK)), only a small study area was selected.

In some case studies, similar water bodies were grouped according to the extent of the main use (e.g. groups of heavily urbanised, semiurban and rural sections) and physical alteration (e.g. groups of natural stretches, reservoirs, regulated river stretches). Grouping of water bodies was used to simplify the further process of classifying ecological status or identifying restoration measures. For instance, in the case study on the River Dee (UK), which is 78.5 km long, a total of forty-two water bodies were identified (mean size of 21 km^2), some of them being very small units of ca. 200 m. As a pragmatic solution to undertaking the ecological assessments and subsequent designation process for HMWB, these water bodies were grouped according to the structure from which their modification arises, and finally according to the highest power station of the cascade at which the water collected by the structure(s) is used. Thus, the numerous water bodies formed six water body groups.

As already mentioned, the lakes of the HMWB case studies were identified as distinct water bodies as suggested in Annex II of the WFD, while in both case studies on transitional waters, the estuaries were divided into three water bodies. For coastal areas, it was not possible to clearly identify separate water bodies.

4.3.2 Scale of Water Body Identification

The assessment of the impact of physical alterations on hydromorphology and biology as well as on the classification of ecological status is scale dependent. At the same time, a water body should represent a reasonable management unit for the purpose of water protection, planning of water use and restoration methods. One of the most important aspects of identifying distinct water bodies is therefore finding an appropriate scale, considering ecological principles as well as needs for reasonable management units.

In the support material (guidance papers) agreed by the HMWB WG for conducting the case studies, it was recommended to investigate the extent of impacts in sections of the following scale:

- Length of 1 km or an area greater than 1 km^2 for catchments <1 000 km^2; and
- Length of 10 km or an area greater than 10 km^2 for catchments >1 000 km^2.

This would define the assessment of the overall level of impact upon ecological status (related also to the level of water body identification) in blocks equal to or larger than 1 km^2 and 10 km^2, respectively.

The above recommendations were actually revised in the final guidance document on HMWB and AWB (WFD CIS Guidance Document No. 4 2003) pointing out the following important issues:

- Scaling due to impact assessment changes according to the pressure and impact characteristics, i.e. some pressures have lower thresholds for wide-scale impacts than others;
- Scaling may change according to the water body type and ecosystem susceptibility. Spatial and temporal scale (resolution of impact assessment) should be more precise in such water body types and specific ecosystems that are considered susceptible to pressure.

Rivers

An extensive discussion regarding the scaling issue was performed in the case study report on the Ruhr River (D). According to the authors of this case study, using the scaling of 10 km for catchments larger than 1 000 km^2 at the lower Ruhr River leads to impounded and free flowing sections being included in the same water body. As a consequence, if 50% of these sections were to be identified as HMWB, the closest comparable natural water body would be a natural lake; for the actually free flowing section of the water body, Good Ecological Potential would then be described by a still-water biocoenosis of a shallow lake. To solve this problem, it was suggested that the size of a water body be defined by the length of the dominant pressure, aggregating similar stretches (= subbodies), as shown in Fig. 4.2. Therefore, if the conditions of a water body are not homogeneous, this water body should be divided into sub-bodies characterised by homogeneity of pressures. In this context, the following procedure was proposed for defining homogenous sub-bodies of water:

- If the catchment is of size <1 000 km^2:
  - If the morphologically modified section (e.g. reservoir, impounded lake, canal) extends over more than 0.1 km, it is identified as a single sub-body (max. length = 1 km), beginning with the upper limit of the morphological modification (homogeneity of pressures in the sub-body);
  - If the morphologically modified section (e.g. small impounded lake, canal) extends over less than 0.1 km, a section of 1 km is identified as a natural sub-body (heterogeneity of pressures in the sub-body due to generalisation effects).
- If the catchment is of size ≥1 000 km^2:
  - If the morphologically modified section extends over more than 1 km, it is identified as a single sub-body (max. length = 10 km), beginning with the upper limit of the morphological modification (homogeneity of pressures in the sub-body);
  - If the morphologically modified section extends over less than 1 km, a section of 10 km is identified as a natural sub-body (heterogeneity of pressures in the sub-body due to generalisation effects).

From the case study on the Ruhr River (D), it may therefore be concluded that the scale used for the identification of water bodies should be ecologically relevant.

The UK-EW guideline report (Booker et al 2002a) also provided a discussion on the issues of scaling, water body identification for rivers and its link to impact assessment, and the subsequent provisional identification of HMWB.

The approach adopted for the UK-EW case studies was to divide the river into separate stretches (water bodies). According to the authors of the UK-EW guideline report, however, this approach focused on too small of a scale. It was argued that instead of dividing the river into a series of water bodies each affected by different modifications, it could have been treated together as a single water body (and subsequently a single provisional HMWB) affected by a combination of modifications. For example, the Tame River was divided into a series of different water bodies, which were affected by different modifications (navigation, urban development, industry (gravel mining)). This separation may not have been necessary given that the overall effect on the river was similar.

The authors of the UK-EW guideline report argued that on the one hand, by dividing the river into separate water bodies, the impacts of measures are not assessed in relation to more extensive cumulative effects that may develop over the entire length of a river. In the UK-EW case studies, upstream and downstream effects were taken

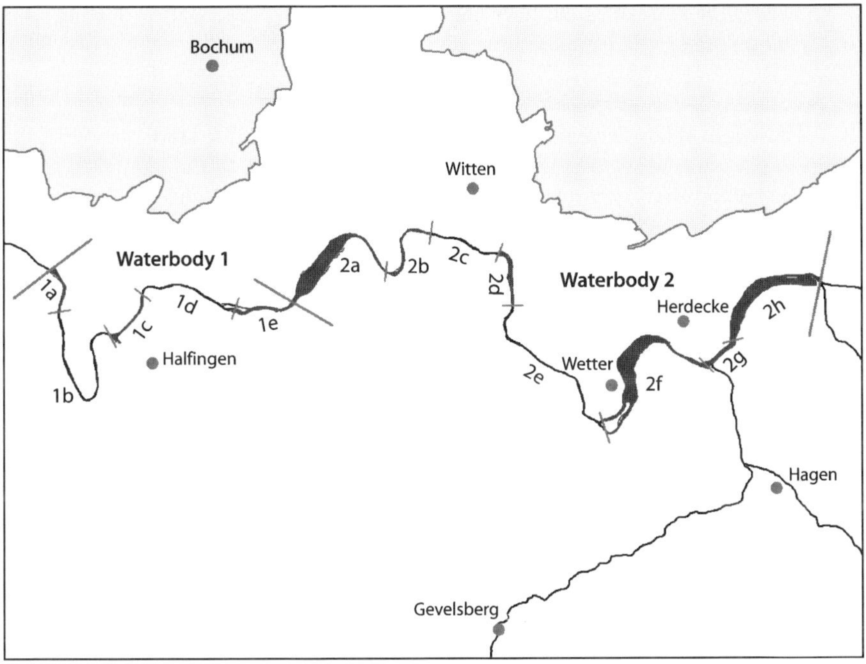

Fig. 4.2. Definition of water bodies and sub-bodies in the lower Ruhr River, Germany. Note: *WB1:* 8.9 km free flowing and 3.3 km stagnant water; *WB2:* 7.9 km free flowing, 5.6 km stagnant water and 14.1 km impounded lakes. *Source:* Case study on the Ruhr River, D

into account, but the extent of such effects was assumed to be small. Overall, this approach could lead to an underestimation of the possible gains that could be achieved through rehabilitation or mitigation, leading to an over-tendency to designate water bodies as being heavily modified.

On the other hand, by not dividing the river into separate water bodies and considering only the likely cumulative effect that the various modifications have on the river, authorities may be forced into achieving "good environmental status" throughout the river, even though this would require adopting measures that are not good value for the money in terms of the costs and benefits.

As a conclusion in the UK-EW guideline report, it was suggested that where river catchments are sub-divided into a number of separate water bodies for HMWB identification and designation purposes, some form of strategic impact assessment is required to ensure that cumulative effects are taken into account. The assessment should be a means of accumulating information on individual water bodies within a catchment in order to provide a more strategic overview of net effects. This would help to ensure that potential synergies (and antagonisms) in the gains that can be delivered by potential measures are identified and examined. Thus, the need for interaction between the HMWB identification and designation process and river basin management planning may be necessary early in the overall decision-making process.

Regarding the scale of water body identification in other case studies dealing with rivers, the following can be summarised:

a Large and very large rivers with catchment >1 000 km^2:
The length of large rivers investigated within the HMWB case studies varied from 70 km (Dender, B) to over 1 000 km (Elbe, D). The length of the identified water bodies ranged from 10 to 490 km with average stretches of 40–90 km. The number of the identified water bodies was usually between three and five.
In the case study on the Dee (UK), a total of forty-two constituent water bodies were identified, equating to a mean size of 21 km^2. However, this average figure involved considerable variation with the smallest units featuring channel reaches of ca. 200 m.

b Small- and medium-sized rivers with catchment <1 000 km^2:
The length of small- and medium-sized rivers investigated within the HMWB case studies ranged from 27 to 80 km. The minimum length of water bodies identified was 1 km, while the maximum length was 54 km encountered in the case study on the Seefelder Aach River (D). The number of water bodies identified was between one and five water bodies with the exception of the Lozoya River (E), where twelve water bodies were identified. In the German case study on the Seefelder Aach, it was indicated that water bodies with a length of >40 km for catchment areas <1 000 km^2 are adequate.

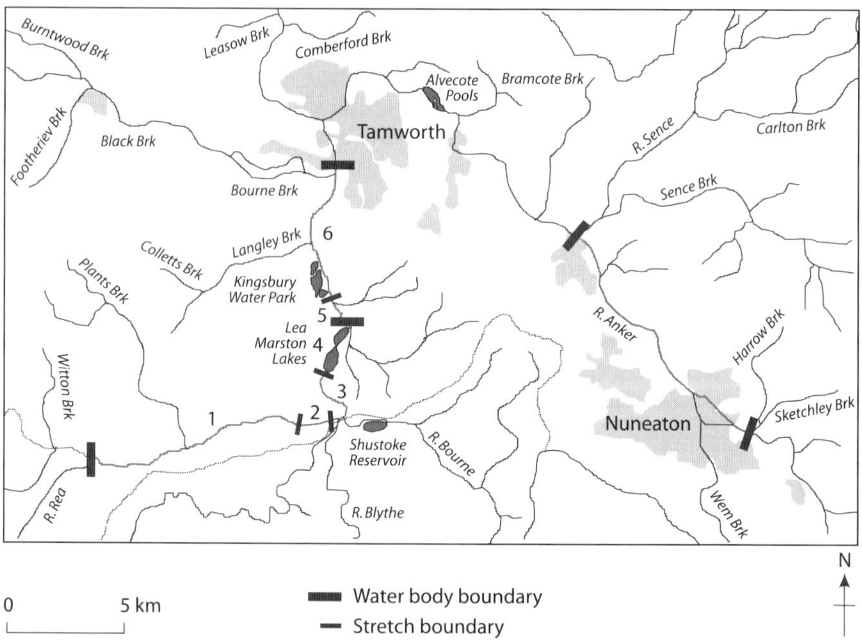

Fig. 4.3. Water bodies and stretches in the case study on the Tame River, UK. *Source:* Case study on the River Tame, UK

The matter of scale was raised in several case studies, especially regarding the minimum size/length of a water body. In the Dutch case study Hagmolen-Hegebeek (total length 27 km), the length of one of the identified water bodies was only 1 250 m. This small size raised the question as to whether the water body is significant enough for distinction and description and it remained unclear what the minimum length of a water body should be.

Two examples of the scale of water bodies identified in small- and medium-sized catchments are shown in Figs. 4.3 and 4.4. Figure 4.3 shows the location of water bodies and stretches in the case study on the Tame River (UK). In this case, one or more stretches are parts of a water body. The stretches are characterised by similar types of engineering and can later be regarded as distinct water bodies. Figure 4.4 shows the location of water bodies in the Solgenan River (lower) sub-catchment, which was the focus of the case study on the Emån River(S).

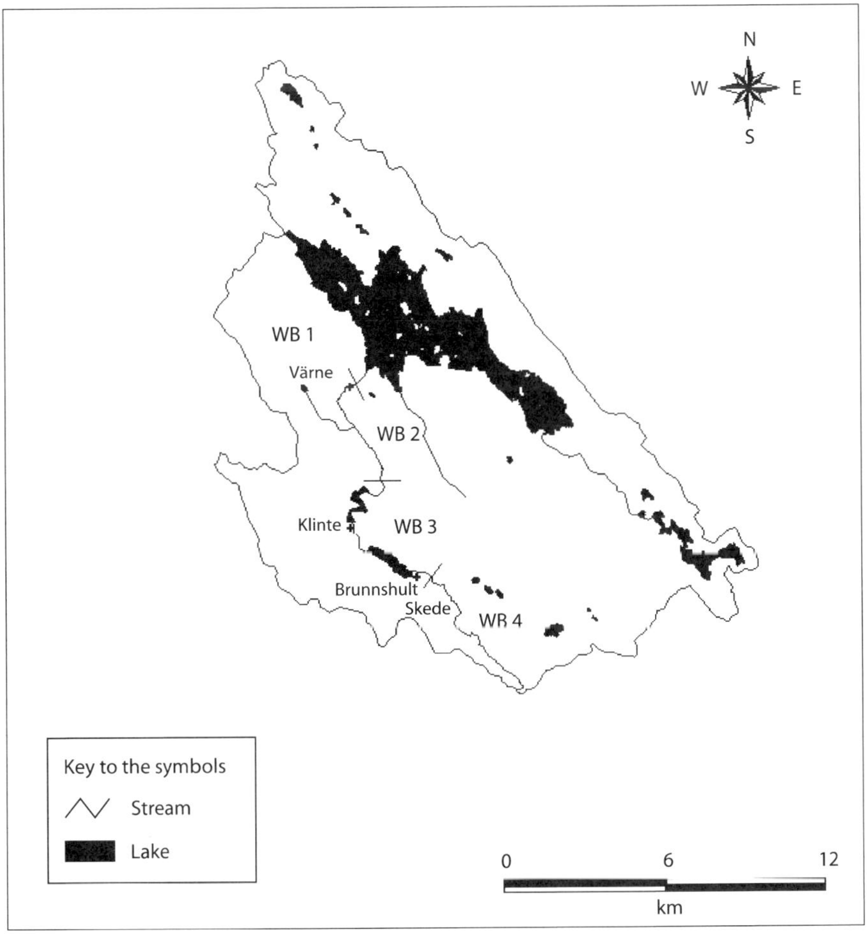

Fig. 4.4. Water bodies in the Solgenan (lower) Sub-Catchment, Sweden. *Source:* Case study on the Emån River, S

Lakes

According to the WFD Annex II 1.2.2 (typology system A for lakes), the smallest size limit based on surface area for lakes is 0.5 km^2. In the Tummel case study (UK) however, it was pointed out that the consideration of the smallest size limit as set by the WFD (0.5 km^2) would exclude twenty-four of the thirty-five lakes of the catchment from being considered as lakes;[1] moreover, eight out of the eleven lakes above the stipulated size threshold were reservoirs and therefore unnaturally large. The authors of the case study concluded that consideration of the lake size typology according to the WFD gives a distorted representation of standing waters for the Tummel River area and by inference for substantial parts of upland Scotland.

Coastal Waters

In the Baltic coast case study (S), no separate water bodies were successfully identified. Nevertheless, it was recommended that a scale of 1 km/1 km^2 is the most suitable for water body identification for such highly variable coastal systems. Along the Swedish Baltic coastlines there are some areas without islands or archipelagos where it could be possible to use the larger scale of 10 km/10 km^2.

Transitional Waters

As far as reported, scale did not play a role in water body identification in the context of the two case studies on transitional waters.

4.4 Differentiation of Artificial Water Bodies

As shown in Fig. 2.1 (p. 12), the second step in the identification and designation process for HMWB and AWB is to identify whether the water body is artificial, i.e. "created by human activity".

According to three HMWB case studies (Lake Loosdrecht (NL), Lake Veluwerandmeren (NL) and Nestos River (GR)), the differentiation of Artificial Water Bodies (AWB) and Heavily Modified Water Bodies (HMWB) remained unclear and required further discussion.

The approach on AWB and HMWB presented in the two case studies on Dutch lakes is presented in Box 4.2.

[1] With regard to the "smallest size limit of the WFD", the reader should also refer to the EC horizontal guidance document on the application of the term "Water Body" in the context of the Water Framework Directive (WFD CIS Guidance Document No. 2 2002). The horizontal guidance document recognises that there are still large numbers of discrete rivers and lakes that are smaller than the WFD thresholds and suggests a possible approach for the protection of these waters. In concrete terms, the document cites the following: "It is recommended to use the size of small rivers and lakes according to system A. However, it is recognised that in some regions where there are many small water bodies, this general approach will need to be adapted. Having said that, it may be appropriate to aggregate water bodies into groups for certain purposes in order to avoid unnecessary administrative burden".

Box 4.2. Differentiation of AWB and HMWB According to the Dutch Case Studies

In the Netherlands, numerous lakes are created by human activity. The Veluwerandmeren is the product of the construction of dikes and the reclamation of a polder, which has turned a former estuary into a freshwater lake chain. With regard to Lake Loosdrecht, the preconditions for the creation of the lake have been created by industrial peat mining. Subsequent wind and wave action eroded the rest of the peat banks, resulting in a system of shallow, interconnected lakes. In both cases, the difference between Artificial Water Bodies and Heavily Modified Water Bodies was defined in the following way:

- In the case where the water body has been dug by humans, the water body is considered artificial (e.g. canals, pools for swimming, pools resulting from mineral extraction). The question remains as to what extent the water body has to be dug (partially or completely) to be artificial;
- In the case where human activity has created the preconditions for creation of the water body but the water body was not dug by human activity, the water body is not considered artificial. It may be natural or HMWB (to be decided after going through the HWMB designation process).

If the criterion of digging of the water body still does not lead to a clear difference in order to be able to differentiate between AWB and HMWB, then a second criterion could be applied: the existence of a natural reference condition on the basis of a comparable water body. If a natural reference condition applies to that water body, it is HMWB. If there is no natural reference condition for that water body, because it differs too much from natural waters, the water body is artificial. On the basis of these two criteria, the Veluwerandmeren and Lake Loosdrecht were identified as Heavily Modified and not Artificial Water Bodies.

Source: Case study on the Veluwerandmeren, NL.

Two of the HMWB case studies (Nestos River (GR) and Lozoya River (E)) dealt with the issue of whether reservoirs should be considered as AWB or not. On the one hand, in the Greek case study on the Nestos River, all reservoirs that had been created by damming a pre-existing river were regarded as Artificial Water Bodies. On the other hand, the large and deep reservoirs in the Spanish case study on the Lozoya River were regarded as a modification of the river and were identified as HMWB and not as AWB. It should be noted that the approach followed in the Spanish case study is in accordance with that of the final guidance document on HMWB and AWB (Sect. 4.3 of the WFD CIS Guidance Document No. 4 2003), while the approach of the Greek case study is not. According to the final guidance document on HMWB and AWB, a water body that has changed category as a result of physical modifications is not an AWB but is considered to be HMWB (e.g. creation of a reservoir due to the damming of a river) (WFD CIS Guidance Document No. 4 2003).

In general, the final guidance document on HMWB and AWB (Sect. 4.3 of the WFD CIS Guidance Document No. 4 2003) defines an Artificial Water Body as a surface water body that has been created in a location where no significant surface water existed before and that has not been created by the direct physical alteration of an existing water body or movement or realignment of an existing water body. According to the guidance document, this does not mean that there was only dry land present before. There may have been minor ponds, tributaries or ditches, which were not regarded as a discrete and significant element of surface water and therefore not identified as a water body.

If the above characterisation of a water body is fulfilled, Member States will have the option to identify them as AWB and consider them for designation or in some circumstances, identify them as natural water bodies. If a Member State considers that Good Ecological Status can be achieved in an AWB, then the Member State may wish to consider the AWB as a natural water body. This would allow Good Ecological Status (GES) to be defined for the water body rather than Good Ecological Potential (GEP).

With regard to the designation of Artificial Water Bodies, the final guidance document on HMWB and AWB (Sect. 5.8 of the WFD CIS Guidance Document No. 4 2003) notes that Member States *may* designate a water body as artificial (as stated in Article 4(3) of the WFD). This suggests that it may not always be necessary to consider designating waters that have been created by man as artificial. There may be some circumstances where long-established water bodies, which are subject to little or no pressures, are indistinguishable from natural waters. Under such circumstances, it may be appropriate to consider their current biological condition as High or Good Ecological Status.

4.5 Conclusions and Discussion

The following are the main findings of the case study evaluation with respect to the identification of water bodies:

- The identification of water bodies forms a crucial component of the HMWB identification and designation process and determines the scale of both technical and management aspects within the process.
- The identification of water bodies should be seen as an iterative process, so that final water body identification may change during the designation process.
- The scale of water bodies differs widely depending on the criteria/approach used for water body identification.
- It was concluded that a water body should be large enough to form a reasonable management unit. For instance, it is reasonable to consider sequential run-off-river impoundments as one water body, although there may be slight differences in ecological conditions.
- Water bodies should not be too small because disproportionate effort would be required for operational monitoring of Heavily Modified Water Bodies.
- Since size and hydromorphological features differ widely, a predefined scale for investigations (e.g. 1 km or 10 km) leading to water bodies of constant size is not appropriate and may not serve the purpose of defining reasonable management units. Rather, it was suggested that scale should be ecologically relevant.
- Different criteria have been used in a different relative order for the identification of water bodies.
- The criterion "effective management units", which takes account of the induced pressures and resulting impacts, was the most important criterion for identifying water bodies in case studies dealing with distinct physical alterations such as dams.
- The criterion "hydromorphological units" was the most important criterion for identifying water bodies in case studies dealing with less distinct physical alterations

resulting from mixed uses such as agriculture, water abstraction or navigation. In such cases of less distinct physical alterations, impacts on the ecological status often occurred gradually. Therefore, it proved difficult to define straightforward changes in impacts between water bodies as the criterion of "effective management units" requires.

- In most cases, the WFD typology systems A and B were not used to identify distinct water bodies. Instead, the assignment of a certain water type to a specific water body was done after water bodies were identified (refer also to relevant discussion in WFD CIS Guidance Document No. 2 2002).
- There was a widespread use of physical features (such as local geology) for water body identification in the case studies.
- Particularly large waters show different patterns of uses along their longitudinal gradient (e.g. case studies on the Lahn and Elbe Rivers, D). In many cases, the headwaters represent widely undisturbed conditions while the downstream sections are more or less modified due to multiple anthropogenic uses. A subdivision according to the significant pressures will lead to the upper, middle and downstream section identified as distinct water bodies. Where appropriate, these sections may be separated in more detail. Small waters with relatively homogeneous pressures do not require a subdivision and can be treated as one water body.
- Although water bodies were identified separately, it was sometimes useful to group them together for assessment and management options. For instance, the approach in the UK-S case studies focusing on hydropower was to group individual water bodies in terms of the hydropower station to which they contribute. In the case study on the Rhone River (F), it was proposed as more efficient to group the numerous canal reaches of the study area into one or two water bodies with similar problems than to consider them separately.
- The need was expressed for clearer guidance on the differentiation of Artificial Water Bodies and Heavily Modified Water Bodies, including the status of Artificial Water Bodies in the overall designation process. As already mentioned, the final guidance document on HMWB and AWB provides some more guidance on these issues (in Sects. 4.3 and 5.8 of the WFD CIS Guidance Document No. 4 2003).

The need for further discussion was identified for the following issues, which remained unresolved during the course of the studies:

- The priority and order of application of the different criteria for the identification of water bodies remained unclear.
- It was also questioned whether administrative borders may be considered for the identification of water bodies. In the German case studies, it was for instance concluded that the subdivision according to administrative entities is not practicable or useful.
- In the case of the Netherlands, it was not clear whether neighbouring lakes, which show a comparable hydrology and pressures but no completely identical ecology, should be considered as different water bodies or rather as one water body.
- It also remained unclear whether it is possible or advisable to use regional approaches and criteria for the identification of water bodies. For instance, the Nordic countries

have suggested using regionally common principles for the identification of water bodies and their size.

- In the case of international rivers such as the case study on the Nestos River (GR), there may be insufficient knowledge of upstream physical alterations. It remained unclear how to best deal with such lack of information.
- With regard to coastal waters, it was concluded that the identification of water bodies can be very complicated. In the case study of Sweden on the Stockholm Archipelago, it remained unclear how to deal with the lack of natural boundaries and lack of knowledge of complex physical and biological patterns.

Chapter 5

Uses, Physical Alterations and Impacts to Hydromorphology

5.1 Introduction

As shown in the identification and designation process for HMWB and AWB (Steps 3 and 4 in Fig. 2.1, p. 12), it has to be assessed whether there are changes in the hydromorphology of water bodies, and subsequently significant changes should be described. Significant anthropogenic pressures (for HMWB and AWB, physical alterations are of relevance) and the resulting impacts should be further investigated and described (WFD Annex II 1.4) as part of the surface water characterisation required by the WFD Art. 5(1) by December 2004. The characterisation involves the identification and description of the main "specified uses", significant anthropogenic pressures (WFD Annex II 1.4), and significant impacts of these pressures on hydromorphology (WFD Annex II 1.5). In the HMWB case studies, this involved a description and assessment of uses, physical alterations (pressures) and significant impacts upon hydromorphology.

This chapter focuses on the methods and approaches used in the HMWB case studies to describe and evaluate the significance of specified uses and physical pressures (physical alterations) on the water bodies as well as the impacts of these pressures on hydromorphology.

In this book, the following definitions of the terms "specified use", "pressure", "physical alteration", and "impact" have been used as given in the final HMWB and AWB guidance document (WFD CIS Guidance Document No. 4 2003):

- "Specified uses" are the water uses described in WFD Article 4(3)(a)(ii)–(v), i.e.,
  - Navigation, including port facilities, or recreation;
  - Activities for the purposes of which water is stored, such as drinking-water supply, power generation or irrigation;
  - Water regulation, flood protection, land drainage; or
  - Other equally important sustainable development activities.
- "Pressure" is a follow up of an anthropogenic activity (specified use),[1] which could impact a water body. The WFD Annex II, requires the consideration, in particular, of pollutants from point and diffuse sources, alterations of the water regime (water abstraction, water flow regulation), alterations of the morphology of the water bodies and of other human activities that may have an impact;

[1] Because of the wording of the WFD and in particular Article 4(3), the HMWB Working Group uses the term "specified uses" and not the term "driving forces" of the IMPRESS Working Group guidance (WFD CIS Guidance Document No. 3 2002).

- A pressure is a direct or indirect consequence of the specified use, defined in terms of the altered environmental variable(s) to which water bodies and their receptor(s) are exposed. This could be a physical structure that is required to maintain the specified use.
- A "physical alteration" is defined as modification of the hydromorphology of a water body by human activity. Essentially, physical alterations are a type of pressure.
- The concept of physical alteration was regarded in the HMWB case studies as very broad; many different types of changes were regarded as falling under this term. As defined in the case studies of the UK-EW, a physical alteration/modification is the change (or changes) made to the river that may be responsible for the river failing to meet Good Ecological Status. Each alteration should serve an intended use (such as straightening for navigation or construction of flood banks for flood defence). Some rivers, however, can have alterations that do not serve the originally intended use any longer (e.g. weirs used to maintain water levels for mills that are not in use anymore). In the Swedish case study on the Baltic coast, physical alterations were viewed as any type of action that changes the morphology or hydrology of the shore, bottom or the open water body.
- An "impact" in the context of the WFD (Annex II 1.5 of the WFD) is a change to the value of the quality elements resulting from one or a number of pressures, which potentially could lead to failing the environmental objectives set in Article 4.
- For the purposes of the HMWB and AWB identification and designation process, significant impacts on hydromorphology are of relevance. According to Annex V (1.1) of the WFD, the elements to be considered are: *river continuity*, *hydrological regime* (quantity and dynamics of water flow, connection to groundwater bodies, residence time in lakes), *morphological conditions* (width variation of rivers, depth variation, structure and substrate of the bed, structure of the riparian zone, of the lake shore and of the intertidal zone) *and tidal regime* (freshwater flow of transitional waters, direction of dominant current of coastal waters and wave exposure for transitional and coastal waters).

5.2 Methods for the Description of Specified Uses

The majority of the HMWB case studies used no particular methodology to describe specified uses. The specified uses were simply qualitatively described. In most cases, specified uses were presented in combination with their related physical alterations and often described for each of the identified water bodies, as in the case on the Baltic coast (S), Forth Estuary (UK) and the German case studies.

Elements of quantitative description appeared in the cases on the Emån River (S) as well as the Dee River (UK) where the percentages of the catchment influenced by different uses were calculated.

5.3 Methods for the Description of Physical Alterations

The methods used in the HMWB case studies to describe the physical alterations (or pressures) ranged from very qualitative or simple descriptive ways of reporting to detailed methods of assessment of the degree of physical alterations. Some of the methods presented in this section have also been used for the assessment of hydromorphological impacts. In these cases, the methods here are not repeated in Sect. 5.4 of this book.

In the majority of cases that used qualitative description, physical alterations (or pressures) were described in relation to the main identified specified uses or/and in relation to the identified water bodies. Some simple parameters and methods used to assess the presence of physical alterations were:

- Comparison of water structural characteristics between past and present (e.g. length of river section, width, width variation, number of islands, number of groynes) (Elbe, D);
- Number of weirs per km and their main characteristics as well as an indication of whether they are passable or impassable for fish (Lahn, D);
- Map presentation of the extent of direct human impacts. In the case of the Forth Estuary (UK), the extent of physical alterations and uses was given as a percentage of both the intertidal and the subtidal area in each water body. The data used were derived by spatial analysis of digital data.

Member States often separated physical alterations (or pressures) into different categories (direct/indirect). In the case study of Hagmolenbeek-Hegebeek (NL), physical alterations were classified into changes to morphology (direct), changes in the catchment area and changes in catchment hydrology (indirect). A distinction between direct and indirect physical alterations was also made in the case studies on Lake Loosdrecht and Veluwerandmeren (NL). Direct physical alterations included embankments, artificial islands, etc., while indirect ones included land reclamation, land drainage, and water inlets. Similarly, in the case study on the Rhone River (F), two types of physical alterations were identified according to the way they affect the environment: (1) direct ones, which can be ascribed to a given activity but were often started in former times (e.g. embanking, river channelling for navigation, which is now done in branch channels) and (2) indirect ones, which often cannot be correlated to one specific activity. An example is the lowering of the longitudinal profile noted in the Rhone, which signals a deficit in solid transport linked to changes in upstream sectors (occupation of catchment basin lands, water works, and stabilisation of formerly mobile beds).

More detailed methods for the description and the assessment of the degree of physical alterations were the following:

- The German case studies (the rivers Elbe, Seefelder Aach, Lahn, Mulde) used the *LAWA (Länderarbeitsgemeinschaft Wasser) criteria*[2] for the identification of significant pressures (or physical alterations) on surface waters. These significance criteria consisted of a mixture of morphological and hydrological parameters and were

[2] Compiled in the context of the LAWA-Subcommittee "Surface waters and coastal waters" following the WFD Annex II. LAWA is the Working Group of the Federal States on Water Problems in Germany.

examined in a modified form in the German HMWB case studies. Individual criteria are provided for pressures caused by the specified uses: navigation, flood protection, hydropower, agriculture, water supply, and urbanisation (see Table 14.10 in Annex II). The assessment according to the LAWA criteria is not carried out for the entire catchment but for the individual water bodies. As a result, the considered scale affects the relative significance of the pressures in the different water bodies. In some cases, the consideration of the entire catchment may provide different results; for instance, specified uses such as flood protection and urbanisation, which have been assessed as insignificant for individual water bodies, may become relevant when considered on a broader catchment scale (e.g. flood retention basins are not among the significant criteria for the assessment of flood protection). This point is particularly important when we consider measures for flood defence where the restoration options may be more sustainable flood management approaches (e.g. wetlands) further up the catchment rather than restoration within the water body.

- The "German Stream Habitat Survey" was used to assess physical alterations in each identified water body. The methodology consists of the mapping of structural aspects of the water body. A predefined system of single parameters is used to classify water bodies into seven classes of morphological status (from pristine to extremely modified) (LAWA 2000; LUA NRW 2001). Pristine condition (class 1) is the present-day potential natural state after abandoning all existing uses and removing structures. Good Ecological Status was equated to class 2 of the Stream Habitat Survey in order to comply with the classification of the WFD. The results are shown in maps of structural water quality.
- In the case study on the Baltic coast (S), the "Indicator Method" was proposed as a method to describe the degree and extent of physical alterations. Its use was recommended especially for recreational developments, harbours etc. and areas with a large number of geographically small disturbances. The method does not measure actual levels of disturbances but the amount or number of potentially disturbing factors. Indicators used are the number of jetties per km of shoreline, number of buildings per km of shoreline, and road length per km of shoreline (using official digital maps, data from aerial photographs, and digital ortho-photographs). Levels of disturbance are classified into five classes (nil, slight, significant, extensive, and very extensive). Results are presented as a map with different colours for the different classes. The results of the method can be used as an indirect assessment of hydromorphological impacts. For instance, the number of jetties can be used for the assessment of the extent of shoreline alteration.
- In the Belgian case study (Dender River), three groups of indicators were used to assess physical alterations to the river: indicators of hydrological quality, stream continuity and stream morphology. To evaluate the intensity of the physical alteration, stress factors were determined in association with the different uses of the water body. For each of these stress factors, indicators were determined (relating to hydrological regime, river continuity, morphological elements of channel, erosion and sedimentation, navigation, and riparian zone). These indicators were evaluated by classification according to the intensity of alterations. Based on the value of each indicator, an order of ranking was set, ranging from "very heavily modified" (class 5) up to "no modifications" (class 0), and the relationship between the degree of modification and the value of each indicator was given in a table (see Table 14.11 in

Annex II). The method was then applied to each of the identified water bodies. The different indicator values were added up in order to be able to evaluate the degree of modification for each water body. The maximum value a section can obtain is the number of indicators multiplied by the maximum value of each indicator. The sum of the indicator values was rescaled up to a maximum of 100 to compare with other studies or scripts. In that way, it is possible to express the degree of modification as a percentage. The following (objective) classification was used to determine the degree of physical modification:

- Sum of the different indicator values >80 = very heavy modification;
- Sum of the different indicator values between 60 and 80 = heavy modification;
- Sum of the different indicator values between 40 and 60 = mediocre modification;
- Sum of the different indicator values between 20 and 40 = limited modification;
- Sum of the different indicator values <20 = very limited modification.

This evaluation method can be used for other streams and catchment areas (if necessary, adjustments of the class limits can be made and other stress factors like abstraction of drinking water can be included). It may be necessary to perform a field survey before using this method in other catchment areas

- In the case study on the Dalälven River (S), seven variables were used to describe the degree of physical modification (see Table 14.12 in Annex II). An index was calculated by adding the values of all indicators.

  In the particular case study, the large size and complexity including the large number of dams made it difficult to gain an overview of the effects of hydropower and other physical modifications. However, the method described above provided an interesting and valuable quantification of the degree of physical modification as measured from GIS data. Moreover, to gain the maximum relevance for investigating effects of physical modification on fish or other organisms, these GIS-based data could be developed or further complemented to match specific problems concerning the group of organisms in focus. For example, the degree of isolation measured as the number of dams from investigated localities to the sea could be obtained.

- The case studies of UK-EW used the so-called "pre-screening methodology" as described in Box 5.1 and Fig. 5.1 to reach a decision of whether a stretch is a potential HMWB or not (see the final step in Fig. 5.1). This was in order to screen out water bodies that are obviously not heavily modified.

  The outcome of the pre-screening process is based on the presence of physical alterations and hydromorphological changes and is subsequently complemented by conclusions on the ecological status of the water body in order to proceed to the provisional identification of HMWB or not. Within the pre-screening process, a detailed description of physical alterations using data from the well-established River Habitat Survey (RHS) method and the Flood Defence Management System (FDMS) is performed. Several parameters measured in relation to both channel and bank modifications were used to derive habitat quality scores as well as Habitat Modification Scores (HMS)[3] within the RHS methodology by applying a simple set of rules to the

[3] At the time of writing, the HMS was currently being reviewed in the light of the wider requirements of the WFD. There were several areas which need fine-tuning for the purposes of HMWB definition, including how unnatural channel width can be scored and scores for culverts and raised water levels (e.g. due to downstream weirs).

RHS data. The HMS was calculated for each General Quality Assessment (GQA) stretch (an identified water body may consist of more than one GQA stretch). However, it was also suggested that at the early stages of an investigation, simple data sources can be used to assess physical alterations, such as Ordinance Survey maps. The ranges of the Habitat Modification Score were classified to correspond to five categories (Habitat Modification Index) as regards the physical state of the river. The RHS data was complemented by the Flood Defence Management System data (flood defence asset information) and local knowledge for a more complete description of the physical alterations.

Box 5.1. Pre-Screening of Water Bodies, UK-EW

To avoid the need to apply the full designation process to all water bodies, it is highly desirable that many water bodies can be excluded from the process at an early stage. A pre-screening methodology is required, which enables exclusions to be established. The approach of the UK (England and Wales) involved the collation of data and the screening out of water bodies that are obviously not heavily modified. The proposed procedure makes use of a combination of available data and analysis of the data by people with local knowledge of the water body. They are well placed to interpret the data, add further details extracted from maps or photographs, and amplify the data using their local knowledge.

The two key sources of data are the River Habitat Survey (RHS) and Agency Flood Defence asset data, the latter generally held in the Flood Defence Management System (FDMS). The data are recorded in a table, and a series of rules are applied to the data. If the rules are satisfied, the water body is not a Potential HMWB and further analysis can be avoided. If the rules are not satisfied, then the water body is a Potential HMWB and further analysis, including application of the designation tests will be necessary to determine if the water body should advance to a designation as an HMWB.

The final pre-screening methodology proposed is not totally objective but requires local knowledge and data together with a degree of judgment. The given table for pre-screening should ideally be completed in a small workshop where staff from different disciplines can add a variety of experience and knowledge to the process.

When the proposed procedure was implemented, each water body was divided into General Quality Assessment (GQA) stretches. Each stretch was assessed using the methodology summarised in Fig. 5.1. Experience and judgment together should be adequate for the assessment of the stretch. If there is any doubt, the stretch should be classed as borderline.

The following guidelines for classification may be useful:

- The presence of high percentages of stretch length with additional flood channels is likely to prevent the stretch being classified as non-candidate;
- The presence of weirs and sluices at a frequency of more than 0.3 per km is likely to prevent the stretch being classified as non-candidate, particularly where weirs or sluices have a major influence on water levels for a long distance upstream of the structure;
- The presence of re-aligned or straightened channel for more than 25% of the stretch length is likely to prevent the stretch being classified as non-candidate;
- If regular maintenance is carried out along most of the stretch, it is unlikely to be classified as non-candidate.

The decision on progression to designation does not relate to each individual stretch, but to the whole water body. If only one stretch is a non-candidate, it is likely that the water body should proceed to the designation tests. If, however, there is a wide disparity in results across the stretches, consideration should be given to changing the boundary of the water body or to subdividing the water body, as it is likely that it is non-homogeneous.

Source: UK-EW guideline report (Booker et al. 2002a).

- Similar to the UK-EW approach outlined above, an initial sorting operation of the identified water bodies was carried out in the case study on the Sarre River (F). This procedure was named selection of pre-candidate water bodies. The purpose of HMWB pre-candidate selection is to investigate the physical characteristics of the

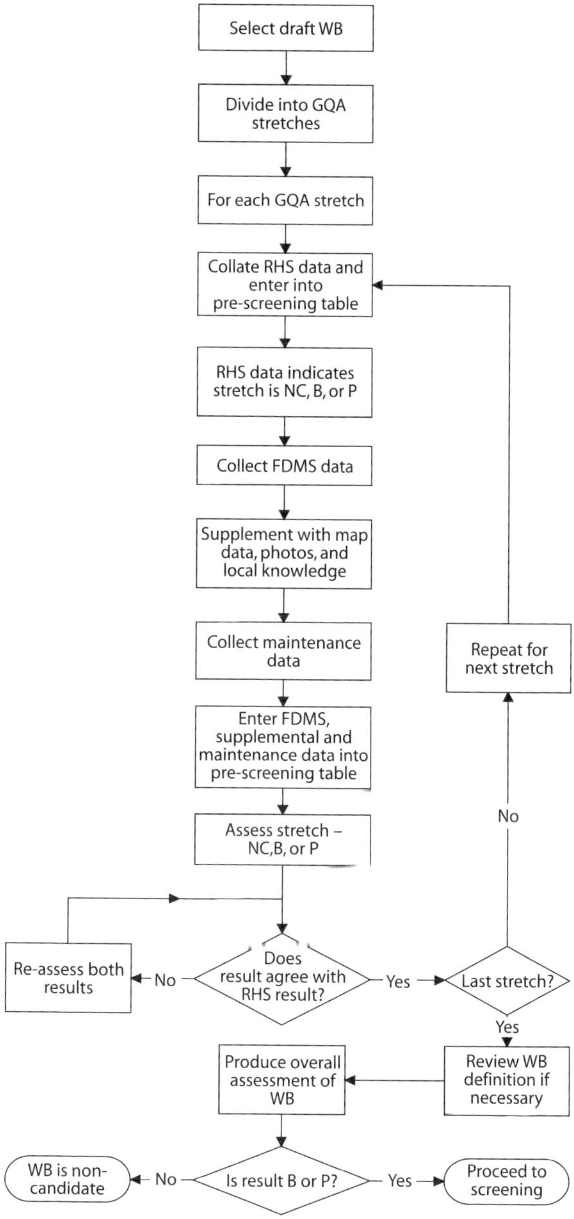

Fig. 5.1. Outline of pre-screening process, UK-EW. Note: *NC* = non candidate, *B* = borderline, *P* = potential. *Source:* UK-EW guideline report (Booker et al. 2002a)

river basin district and to make a first selection eliminating "obviously little modified and easily reversible" water bodies. The method involved the identification of criteria to evaluate the degree of modification.

In the approach that followed, the physical characterisation was carried out using transparent and replicable indicators for the description of the water bodies. The following indicators (control variables) have been proposed: stream discharge, sediment discharge, riparian vegetation and valley gradient. These provide a brief physical description and a characterisation of modification. The control variables are subject to disruptions caused by environmental modifications as shown in Table 14.13 in Annex II. The degree to which the control variables have been modified was estimated using the response variables available (evolution of the fluvial style, mean flow) and accurately documented for each water body in *diagnostic datasheets*, which were used for the description and classification of each section. In order to physically describe the individual sections and their modifications, indicators appropriate to each case were proposed (e.g. proportion of riparian vegetation, status of riparian vegetation and bed shading for the control variable "riparian vegetation") and the most damaging modifications were identified (e.g. disturbance of hydrology, flow rate control, disruption of the flood expansion field, and artificial modification of stream channel mobility in the case of the Sarre).

The assessment is based on expert opinion and must answer the question: "Has the section been significantly modified, and if so, has it been done in a neither obviously reversible nor easily reversible way?"

The question above can be broken down into four intermediate questions:

- Q1a – Has the stream discharge been heavily modified, and if so, has it been done in a neither obviously reversible nor easily reversible way?
- Q1b – Has the sediment discharge been heavily modified, and if so, has it been done in a neither obviously reversible nor easily reversible way?
- Q1c – Has the riparian vegetation been heavily modified, and if so, has it been done in a neither obviously reversible nor easily reversible way?
- Q1d – Has the valley gradient been heavily modified, and if so, has it been done in a neither obviously reversible nor easily reversible way?

An affirmative answer to just one of these four questions was enough to designate the section as a pre-candidate HMWB, and an ecological analysis was then carried out. The decision criteria were determined and justified by the expert responsible for selecting the pre-candidate water bodies. As the only purpose of this preselection phase is to decrease the workload in the subsequent stages of the process, any doubt should result in the water body being designated pre-candidate.

As a result of the above discussions in the HMWB case studies, the final greed process on the identification and designation of HMWB and AWB included a screening process (Step 3 of Fig. 2.1, p. 12). The screening step aims to reduce effort and time in identifying water bodies that should not be considered for the HMWB designation tests. Screening includes those water bodies that are likely to fail to achieve Good Ecological Status (GES) but show no hydromorphological changes. According to the final guidance document on HMWB and AWB (Sect. 4.5 of the WFD CIS Guidance Document No. 4 2003), for those water bodies which have not been "screened out", significant anthropogenic pressures and the resulting impacts should be further investigated and described (WFD Annex II 1.4).

In the context of the HMWB and AWB identification and designation process, changes to hydromorphology resulting from "physical alterations" are relevant (WFD Art. 2(9)). According to the final guidance, physical alterations include alterations in the morphology and hydrology of the water regime. Physical alterations should usually serve a specified use, such as straightening for the purpose of navigation. However, physical alterations, which do not serve a particular specified use any longer, should also be identified and described in the characterisation (e.g. weirs used to maintain water levels for mills that are no longer in use).

For the characterisation it is important to find out which pressures are of "significance", because only significant pressures (or physical alterations) are to be considered. Member States may use qualitative or quantitative approaches to describe the degree and level of significance of the physical alterations.

5.4 Methods for the Evaluation of Impacts upon Hydromorphology

For the purpose of this document, methods used to describe changes and impacts upon hydrology have been separated from methods used for impacts upon morphology.

Evaluation of Impacts upon Hydrology

The majority of case studies made use of simple parameters and appraisal techniques, both qualitative and quantitative, in order to assess changes and impacts upon the hydrology of the water bodies. In general, a quantitative description of hydrology was considered necessary to enable a better assessment of the hydrological impact on ecological conditions. Examples of parameters used in rivers were:

- Changes of the groundwater table;
- Indicators of water level regulation such as annual and summer water level fluctuation;
- Change in maximum and minimum river discharge before and after construction of a storage reservoir;
- Comparison of annual mean flows and percentage of glacier cover under natural and regulated conditions;
- Quantitative comparison of discharge data with natural seasonal discharge patterns;
- Explicit use of the WFD elements and qualitative assessment: quantity of flow (assessed simply as running, stagnant or dry), dynamics of flow (assessed in terms of presence or absence in variation of flow).

A relevant example of hydrographs used in the case study on the Beiarelva River (NO) is given in Fig. 5.2.

The UK-S case studies used the *Dundee Hydrological Regime Alteration Method (DHRAM score)* to assess the severity of physical alterations to the hydrological re-

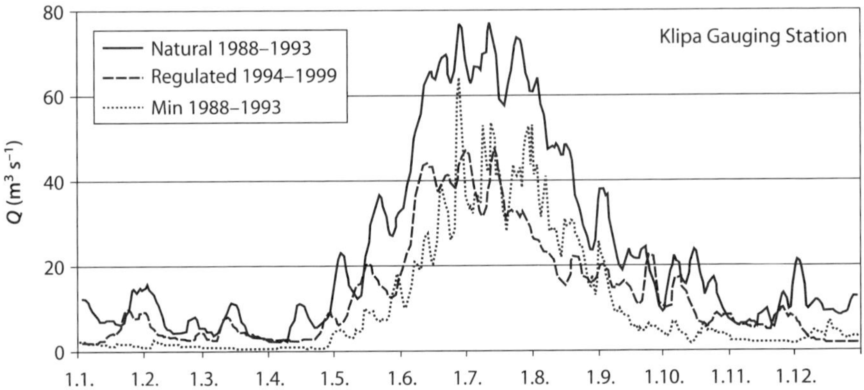

Fig. 5.2. Hydrographs showing hydromorphological change, Norway. *Source:* Case study on the Beiarelva River, NO

gime of rivers. The score ranges from 1 (natural) to 5 (severely impacted), while scores 2 and 3 are considered compatible with Good Ecological Status (GES). The DHRAM scores are calculated for individual water bodies. The method uses empirical and synthetic data to describe mean daily river flows under natural and impacted conditions. According to the case study authors, the DHRAM approach is derivative of widely cited research (cf. Poff et al. 1997; Richter et al. 1996, 1998). Nevertheless, the case study authors identified that important calibration and validation experiments remain to be completed on the DHRAM outputs.

With regard to lakes, in the Finnish case study (Lake Kemijärvi), the changes in water levels were analysed applying the *REGCEL water level analysis.* REGCEL is the main tool used in assessing the impacts of water level regulation in Finnish lakes. The REGCEL water level analysis was also applied in the Finnish case study to provisionally assess the impacts of current regulation on biological quality elements. The analysis consisted of the following phases:

- Determination of regulated and non-regulated (re-calculated values) daily water levels from 1980 to 1999;
- Identification of the indicators describing the effects of water level fluctuation on the littoral environment, fish, birds and recreational use;
- Identification of the most critical water level changes affecting the status and use of water course.

The indicators used were divided into general and specific ones. General indicators of regulation were annual water level fluctuation, water level fluctuation during the summer and water level decrease during the ice-covered period. The specific indicators depicted the impacts of regulation on frost- and ice-sensitive macrophytes, benthic invertebrate fauna species as well as the reproduction of pike.

In the case study on the Haringvliet Estuary (NL), simple parameters were used and their values before and after the closure of the Haringvliet Dam were compared.

The parameters used included salt intrusion patterns, maximum current velocity and tidal volume, mean high water level and mean low water level as well as tidal amplitude.

Evaluation of Impacts upon Morphology

The case study on the Dender River (B) used a series of parameters to assess the morphological changes and resulting impacts. These were:

- Graphical comparison of the natural and artificial linear profile;
- Use of sinuosity to determine the extent to which the stream winds;
- Evaluation of the structural quality of the banks by combining the factors of sinuosity, land use, presence of dikes on the basis of field surveys and examination of orthographic maps;
- Width and depth ratio;
- Sediment delivery ratio.

The Spanish case study (Lozoya River) made use of the QBR Index (Q = Qualitat (Quality); B = Bosc (Forest); R = Ribera (Riparian)) to assess the impact of hydromorphological changes on the riparian zone of sections without reservoirs. QBR is an index that evaluates the quality of riparian ecosystems in Mediterranean rivers (Munné et al. 1998, 2003). The index ranges between 0 and 100 and takes into account the following four components of riparian habitat: (1) total cover of riparian vegetation; (2) cover structure; (3) cover quality; and (4) channel alterations. It classifies results into five classes of riparian habitat quality (high, good, fair, poor, and very poor). According to this method, the quality of riparian vegetation is highly dependent on the structural integrity of streambed and flow conditions.

In another case (Hagmolenbeek River, NL), the WFD elements for morphological status were explicitly used in combination with their qualitative description. In specific, the elements applied included river depth and width variation (assessed as the presence or absence of a natural meandering pattern), structure of the riparian zone (natural or semi-natural vegetation), structure and substrate streambed (qualitative description of presence of silt, sand, gravel, organic matter and dead wood). It was concluded that qualitative description of morphological impacts is sufficient for determining the main problems, whereas for hydrological impacts more quantitative assessment is required.

Other parameters used for the determination of morphological impacts and changes *due to hydropower* were:

- Comparison of the present shape of hydropower reservoirs with their "pre-hydro" shapes before closure of the dam;
- Calculation of % change of area and perimeter of the reservoirs between the pre-hydro and present situation. Changes are considered as significant in terms of impact on biota when the perimeter has increased by 20% and the surface area has doubled.

According to the final guidance document on HMWB and AWB, both qualitative and quantitative appraisal techniques can be used for assessing impacts on hydromorphology resulting from physical alterations. The elements examined should include the elements required by the WFD (WFD Annex V 1.1: river continuity, hydrological regime, morphological conditions, tidal regime), as far as data are available (WFD CIS Guidance Document No. 4 2003).

5.5 Results on Specified Uses and Physical Alterations

The main specified uses identified in the case studies were navigation, flood protection, hydropower, water supply, agriculture/forestry, urbanisation as well as industry and recreation (Table 5.1). Additional specified uses that were identified in only a few case studies were commercial fishing, fish farms and agricultural as well as industrial land take.

The specified uses are briefly described below, including information on the main physical alterations associated with them in the HMWB case studies. The relevant information and results from the case studies have been summarised in Table 14.4–14.9 (see Annex II). For the purpose of description, specified uses with similar physical alterations and resulting impacts have been grouped, i.e. hydropower is presented together with water supply, agriculture together with forestry and livestock (and fish farms), and urbanisation together with industry.

Navigation

Navigation was identified as a specified use in seventeen case studies, in six as a specified use of high intensity, in seven as a specified use of intermediate intensity and in four as a specified use of lower intensity.

For the purpose of navigation, several modifications were made to the river channel and bed of the case study areas. At some places, the rivers were straightened and canalised. Locks, weirs and sluices were constructed. Physical alterations to the river banks included bank reinforcement, dikes, culverts as well as floating quays and marinas for the use of navigation.

Similarly, shoreline reinforcement and shipping channels altered the physical status of transitional waters. Physical alterations in lakes identified in the case studies included shipping channels, deepening of shallow waters and creation of islands. Shipping channels also affected the coastal areas considered in the case study on the Baltic coast (S) along with harbours and landing stages on the shoreline.

Flood/Coastal Protection

Flood/coastal protection was identified as a specified use in twenty-seven case studies, in nine as a specified use of high intensity, in fifteen as a specified use of intermediate intensity, and in three as a specified use of lower intensity.

To protect the environment against flooding, several modifications were made to river channels including, among others, flood protection dams, straightening/channelisation and weirs. Alterations to the river banks included bank reinforcement, embankments and dikes. The shoreline of transitional and coastal waters is also altered by the presence of reinforcements/sea walls and straightening of the shoreline.

Hydropower/Water Supply

Hydropower was identified as a specified use in nineteen case studies, in fifteen as a specified use of high intensity and in four as a specified use of intermediate intensity. Water supply was identified as a specified use in eleven case studies, in three as a specified use of high intensity, in three as a specified use of intermediate intensity, and in five as a specified use of lower intensity.

Among the most frequent physical alterations to the river channel/bed for the purpose of hydropower production were dams, storage reservoirs and impoundments as well as water diverting structures, pipelines and aqueducts. In the case study of the Umealven River (S), a distinction was given for two types of dams:

- Dams built to provide storage reservoirs where water from periods of high flow can be stored for use in periods of little flow to allow for "steady" power production. The dams are built at the outlets of lakes. As a result, lake levels are often dammed to higher levels, thus flooding large land areas surrounding the lakes.
- Dams built to control the flow to hydropower stations, damming so-called run-off-river impoundments. These are located in rapids, where the river loses a lot of its fall height over a short distance. These dams generally flood less land areas compared to the storage reservoirs, but significant portions of the river valley close to the river are affected (see Fig. 5.3).

Physical alterations to the river banks connected to hydropower included bank reinforcement and removal of riparian forest.

Agriculture/Forestry

Agriculture/forestry was identified as a specified use in thirty case studies, in five as a specified use of high intensity, in twelve as a specified use of intermediate intensity and in thirteen as a specified use of lower intensity. Apart from the physical alterations resulting from agriculture and forestry itself, impacts from livestock and fish farms can also be classified under this "specified use" category.

Examples of physical alteration to river channels resulting from agriculture were diversion dams and weirs. Weirs are used to reduce periods of low flow discharges in the summer. Other physical alterations mainly affected river banks and the riparian zone and included land drainage and network of ditches, artificial channels that are parallel and subsidiary to the main river course, land reclamation as well as constructions such as cattle fences for livestock. Probably the most significant physical alteration associated to agriculture in the HMWB case studies was land drainage, which is undertaken to remove water from the land. Land drainage aims to achieve faster precipitation discharges and optimal water levels for agricultural purposes. Much of the fertile, low-lying agricultural land is extensively drained by networks of artificial channels. The effects of land drainage on the hydrological regime are complex and may lead to an increase or decrease of downstream flood peaks (Robinson 1990).

Physical alterations connected to forestry activities included the construction of timber floating structures in the river channel. Excavations and clearings are often performed with the intention of adjusting the river channel for timber floating. The

Table 5.1. Specified uses in the HMWB case studies. *Source:* Compiled by the authors from the evaluation of the HMWB case studies

Country	Name of case study	Navigation	Flood/coastal protection	Hydropower	Water supply	Agriculture/ forestry	Urbanisation	Industry	Recreation	Other specified uses
A	Bregenzerach R.	–	–	+++	–	–	–	–	–	–
	Danube R.	++	+++	+++	–	–	–	++	–	–
	Wienfluss	–	+++	–	–	–	++	–	–	–
B	Dender R.	+++	+	–	–	+	+	++	–	–
D	Elbe R.	+++	+++	–	–	++	+	–	+	Fishing
	Seefelder Aach R.	–	++	+++	–	+++	–	+	+	–
	Lahn R.	++	++	+++	–	+	+	+	–	–
	Ruhr R.	+	++	++	+++	+	+++	++	+++	–
	Mulde R.	–	++	+++	++	++	++	++	–	–
	Dhünn R.	–	++	++	+++	+	+++	–	++	–
E	Lozoya R.	–	–	++	+++	+	+	–	+	–
SF	Kemijärvi L.	–	++	+++	–	+	+	+	+	Fish farms
F	Authie R.	–	++	++	–	+++	++	+	+	Fish farms
	Rhone R.	+++	+++	+++	–	+	++	+	–	–
	Sarre R.	+	++		–	++	+	–	–	–
GR	Nestos R.	–	++	+++	–	++	–	–	–	–
NL	Haringvliet Est.	+++	+++	–	++	+	–	–	–	–
	Hagmolenbeek–Hegebeek R.	++	++	–	–	+++	–	–	–	–
	Loosdrecht L.	++	++	–	+	++	+	–	++	Fisheries
	Veluwerandmeren	+	++	–	+	++	+	+	+	Fisheries

Table 5.1. *Continued*

Country	Name of case study	Navigation	Flood/coastal protection	Hydropower	Water supply	Agriculture/ forestry	Urbanisation	Industry	Recreation	Other specified uses
NO	Suldalslagen R.	–	–	+++	–	+	+	+	+	–
	Beiarn R.	–	+	+++	–	+	–	–	+	–
S	Emån R.	–	–	+++	–	+	+	–	–	Fishing
	Dalälven R.	–	–	+++	–	++	–	–	–	–
	Umealven R.	–	+	+++	–	++	+	+	+	Fishing
	Baltic coast	+++	+++	–	–	++	+++	+	+++	Fish farms
UK (EW)	Kennet R. (Thames)	+++	++	–	+	++	++	++	–	–
	Tame R.	++	+++	–	–	+	+++	++	–	–
	Sankey Brook	+	++	–	++	++	+++	++	–	–
	Great Ouse R.	+–	+++	–	+	+++	++	++	–	–
UK (Scot)	Forth Est.	+–	+++	–	–	++	+++	+++	+	Agricultural land take, industrial land take
	Tummel R.	–	–	+++	–	–	–	–	+	–
	Dee R.	–	–	+++	+	+	–	–	+	Fish farms
UK (NI)	Lagan R.	–	++	–	–	+++	+++	+	+	–

Note: +++ = specified use of high intensity, ++ = specified use of intermediate intensity, + = specified use of lower intensity.

a

b

Fig. 5.3. The Umealven River valley at Bjurfors, Sweden; **a** before regulation; **b** after regulation. *Source:* Case study on the Umealven River (S), taken from Grelsson and Nilsson (1982)

effects of forestry were reported as largest on small streams. Ditching of small streams to lower surrounding water tables and increasing tree productivity completely transforms the stream and its ecosystem, including the destruction of fish populations. Moreover, clear-cutting reduces shading of the stream, causing water temperatures to increase and to be more fluctuating, adversely affecting many species of fish and aquatic invertebrates. The main effects of forestry on large rivers, like the Umealven River (S), are the logging of their riparian forests and a reduced input of coarse woody debris to the rivers. In pristine rivers, coarse woody debris disturbs riparian vegetation, creating sites for the establishment of competitively inferior species and providing substrate and shelter for aquatic invertebrates.

Urbanisation/Industry

Urbanisation was identified as a specified use in twenty-four case studies, in seven as a specified use of high intensity, in six as a specified use of intermediate intensity, and in eleven as a specified use of lower intensity. Industry was identified as a specified use in eighteen case studies, in one as a specified use of high intensity, in eight as a specified use of intermediate intensity and in nine as a specified use of lower intensity.

Fig. 5.4. Urbanised section of the Wienfluss, Austria. *Source:* Case study on the Wienfluss, A

Physical alterations to the river channel associated with urbanisation included dams and weirs, straightening, gravel extraction as well as pits and mines associated with the mining industry. Alterations to the river banks included drainage ditches, waterways and channels, bank reinforcement and numerous infrastructures such as buildings, roads and bridges close to the water (see also Fig. 5.4).

Industry and human settlements are associated with water abstraction, which has significant effects on the hydrology of rivers. Abstracted water is used as cooling water for the industry, drinking-water, irrigation, etc. As concluded in the case studies, water abstractions are not characterised as physical alterations, but they can have impacts upon hydromorphology similar to physical structures. The UK-EW guideline report suggested that the hydromorphological changes to be considered are only those resulting from significant physical modifications of the water body associated with a use. Therefore, changes such as some abstractions and discharges that impact hydromorphology but do not involve significant physical modification are not relevant to HMWB. These should be considered under the general requirements for hydromorphology for good water body status.

Recreation

Recreation was identified as a specified use in seventeen case studies, in two as a specified use of high intensity, in two as a use of intermediate intensity, and in thirteen as a use of lower intensity.

Different types of recreational use of water bodies (or banks) like yachting, angling, biking and hiking may occur along rivers, lakes and coastal waters, thus putting pressure on the aquatic ecosystem. These uses may not be relevant unless they require physical modifications to the water body. Physical alterations to the river channel associated with recreation can include shipping channels and the creation of impounded lakes. Most physical alterations resulting from recreation affect the banks and the riparian zone including marinas and tow paths, picnic and recreation areas, land reclamation, and building construction.

5.6 Results on Hydromorphological Impacts

A detailed list of the hydrological and morphological changes/impacts related to each of the main specified uses can be found in Table 14.4–14.9 (see Annex II). Information in these tables has been compiled from the HMWB case studies. Only a selection of the reported impacts and changes will be presented in the next paragraphs (the reader should refer to the tables for more details).

Navigation

Hydromorphological changes and impacts caused by navigation were reported for rivers and coastal waters.

Rivers
Some of the main hydrological changes resulting from navigation-related modifications included disruption of the river continuum, change of the hydrologic regime due to constructions such as weirs and elevated flow velocity due to channel regulation.

Morphological impacts included, among others, the loss of morphological variety in cross- and longitudinal profile, loss of the original river length and decreased sinuosity, disruption of sediment transport, change of bank structural characteristics and reduced exchange between the river and its riparian zone.

The presence of weirs in particular can exacerbate low flows in summer periods and additionally enhance deposition in areas of the river upstream during low flow periods. Conversely, weirs contribute to erosion downstream during periods of high precipitation as well as releasing higher than normal sediment loads under spate conditions. However, weirs can have positive effects as well, such as an increase in the groundwater table, which contributes to wetness of the floodplains and more frequent flooding, thus enhancing naturalness of the riparian zone (case of the Lahn River, D).

Coastal Waters
The presence of quays for boats can interfere with water replacement and cause changes in the flow direction. From a morphological point of view, the associated structures may cause bottom and shore erosion, while boats re-suspend the deposited silt.

Flood Protection

Hydromorphological changes and impacts caused by flood protection were reported for rivers, lakes, transitional waters, and coastal waters.

Rivers
The structures associated with flood protection may cause significant hydrological changes to rivers including loss of vital periodic flooding, elevated flow velocity due to channel regulation and reduced flooding volume. Connectivity from ground water can also be disrupted.

Morphological impacts/changes include mainly changes to the bank structure, which is often uniform and steep, along with loss of the original length of the river, detachment and restriction of natural floodplains, and reduced longitudinal and lateral connectivity.

Lakes

Hydrological changes in lakes occurring in connection with flood protection involved mainly a fixed water level through water level management. This leads to an unnatural lake morphology.

Coastal Waters

Similarly to navigation, the presence of physical alterations in coastal areas can cause changes in the flow direction and bottom and shore erosion.

Transitional Waters

Flood protection structures in transitional waters altered the tidal prism and reduced the tidal volume and amplitude. Morphologically, the sediment exchange with intertidal areas was interrupted and the overall dynamics of sediment transport were altered. Additionally, the intertidal areas were reduced.

Hydropower/Water Supply

Hydromorphological changes and impacts caused by hydropower and water supply were reported for rivers and lakes.

Rivers

The operation of reservoirs resulted in water level changes in the reservoirs and change of the hydrological regime downstream of the reservoir's dam. The flow regime downstream of dams becomes artificial with reduced discharge (drying out in extreme cases). Structures associated with the production of hydropower may cause a river system or estuary to change to a freshwater lake. Extreme peaking amplitudes and reduced flood peaks were also noted as significant hydrological impacts as well as increased ice cover upstream of dams in connection with an overall reduction of glacier coverage in catchments in northern regions.

Morphological impacts included disruption and reduction of sediment transport and general change of sedimentation pattern and change of bank morphology.

According to the case of the Umealven River (S), the presence of consecutive reservoirs makes it difficult to distinguish upstream and downstream impacts.

Lakes

Hydrological changes in lakes included water level regulation with an increase of the lake level fluctuation. The natural water level in lakes is normally raised in the summer and lowered in winter. In Lake Kemijarvi (SF), water level decreased during ice-covered periods, leaving large areas of the lake under ice exposure.

Agriculture/Forestry

Hydromorphological changes and impacts caused by agriculture and forestry were reported for rivers and lakes.

Rivers

The hydrological changes resulting from agricultural use included changes in the hydrological regime of the river, leading to artificial flow and also affecting the connection to groundwater bodies. Land drainage in particular led to a lower groundwater table and a lower groundwater buffer capacity in the case of the Hagmolenbeek-Hegebeek River (NL). Due to the lower ground-water buffer capacity, a dry period leads to low discharges in a shorter time interval. During wet periods, the ditches transport precipitation from the catchment very fast to the stream, causing high discharges.

Morphological impacts included detachment of wetlands, restriction of natural floodplains and change of the erosion/sedimentation pattern with an increased sediment supply from fields into the rivers.

Lakes

A hydrological impact reported in lakes was the reduction of seepage water into the lake due to reclamation of the adjacent land area.

Urbanisation/Industry

Hydromorphological changes and impacts caused by urbanisation and industry were reported for rivers and coastal waters.

Rivers

Hydrological impacts to rivers resulting from urbanisation mainly included altered frequency of flooding and change of the hydraulic characteristics. The latter may involve flashier flow due to reduced infiltration from the riparian zone. Connectivity with ground water and surface water runoff from land can also be disrupted. Morphological impacts included uniform bank structure, change of river depth and width, reduced riverbed diversity and restriction of natural floodplains.

Lakes

Sand extraction, connected to the need for construction material, was in one case the cause for the occurrence of holes on the lake bottom.

Recreation

Hydromorphological changes and impacts caused by recreation were reported for rivers.

Rivers

Although recreation is rarely the main use on a water body, it can have significant impacts on the morphology of a river, mainly affecting the bank structure and quality.

5.7 Conclusions and Discussion

Specified Uses and Physical Alterations

The conclusions on specified uses and physical alterations (or pressures) are summarised as follows:

- The main specified uses identified in the case studies were navigation, flood protection, hydropower, water supply, agriculture/forestry, urbanisation as well as industry. Additional categories of uses were recreation, agricultural land take and industrial land take.
- Different methods were used for assessing the significance of pressures and the extent of physical alterations. Descriptions and assessments were usually qualitative. Some quantitative approaches were also applied such as the methods proposed in the German, Belgian, Swedish and UK-EW case studies.
- Potentially, some Member States have a large number of modified water bodies that need to be assessed for HMWB identification and designation. It may be appropriate to develop screening approaches to screen out early in the process water bodies that are obviously not heavily modified. Such a (pre-)screening process was proposed as part of the UK-EW guideline report making use of available data and local knowledge of the water body. In the case of the Sarre River (F), a selection of pre-candidate water bodies based on the degree of physical modification was also proposed. As a result of these discussions in the HMWB case studies, the final process on the identification and designation of HMWB and AWB included a screening process (Step 3 of Fig. 2.1, p. 12). The screening step aims to reduce effort and time in identifying water bodies that should not be considered for the HMWB designation tests. Screening includes those water bodies that are likely to fail to achieve Good Ecological Status (GES) but show no hydromorphological changes.
- It was often difficult to separate multiple uses and their impacts in the same case study area. It was also reported difficult to differentiate clearly between a pressure, a physical alteration and a hydromorphological impact.

  As stated in the case study on the Baltic coast (S), it is not absolutely clear how various factors should be divided up in order to be able to identify them either as a pressure or as the impacts of a pressure. For instance, the Stockholm archipelago is subject to many different uses, which make separation of specific effects of certain activities impossible. In that area, the impacts of dense population and recreational activities and their associated physical structures in or close to the water are difficult to differentiate.

  In the case study on the Great Ouse (UK), it was also argued that there is often little or no distinction between "land drainage", "agriculture" and "flood defence" pressures in terms of the resulting modifications. A range of pressures may lead to a particular modification, e.g. re-sectioning and dredging can benefit both land drainage for agriculture and flood defence. Also the same pressures can lead to different modifications in different circumstances.
- The concept of physical alteration has been regarded as very wide; many different types of changes can be regarded as falling within this definition. As defined in the case studies of the UK-EW, a physical alteration (modification)[4] is the change (or changes) made to the river, which may be responsible for the river failing to meet Good Ecological Status. Each alteration should serve an intended use (such as straightening for navigation). Some rivers, however, can have alterations that do not

[4] In the HMWB case studies, the term "physical alteration" has often been referred to as "physical modification".

serve the originally intended use any longer (e.g. weirs used to maintain water levels for mills that are not in use anymore).

- Some rivers have been modified over a considerable period of time; therefore, it can sometimes be difficult to detect the modifications that have been made. Assessment methods for historical and less obvious modifications are needed.

  According to the authors of the case study on the Great Ouse (UK), lowland rivers in England have been modified for thousands of years. This may result in underestimating the degree of physical alteration. A river may look more "natural" than is actually the case, and historical records of works may not exist or be easily available. In relation to this, some concerns were noted in the UK-EW guideline report over the ability of the River Habitat Survey data and methodology to score modification adequately, since the method records only obvious modification. Historical modifications, such as widening and dredging of large rivers, may not be noted, although they could result in significant deviations from reference conditions. Therefore, methods for the assessment of less obvious modifications, such as widening, are required.
- *Issue of scale*: According to the authors of the case study on Lake Kemijärvi (SF), the scale for investigation of impacts should be flexible. The scale should depend on the degree of pressure, i.e. in a river where major changes have occurred due to human uses, the approach should be more detailed.
- *Issue of data availability*: Lack of recent data on morphological alterations was expressed in several cases. The need was expressed to develop methods for the assessment of the structural status of rivers building on the experience of other countries (e.g. the UK River Habitat Survey and the German Stream Habitat Survey). At the time of writing, a CEN standard methodology for assessing morphology building on this existing experience was being developed.

Hydromorphological Impacts

A great variety of indicators, parameters and methods were used to evaluate impacts upon hydrology and morphology. The range of the hydromorphological impacts and changes was also wide. The impacts have been summarised according to categories of specified uses as well as categories of water (Table 14.4–14.9 in Annex II).

The main points of discussion, including problems identified, were the following:

- *Data availability on hydromorphological impacts* was not sufficient in many case studies (e.g. Nestos River (GR), Emån River (S)).
- Many case studies also lacked available information on the hydromorphology of the water body before modification (e.g. prior to regulation for the purpose of hydropower production in the Suldalslagen River (NO)). In the case of the Beiarn River (NO), it was also deemed difficult to compare only 4.5 years of flow observations following the diversion scheme with more than 60 years of flow observations prior to the diversion.
- It was suggested in the case studies that more work is required on the *cumulative effect of small physical alterations,* which do not cause a significant impact on their own. For instance, along the Baltic coast (S), there are many small-scale physical structures which are in or close to the water. Individually, these changes do not cause ex-

tensive damage to their surroundings, but with thousands of them coinciding in the same area permanent changes can result.

- Some impacts can be caused by the abandonment of a certain use and the lack of maintenance of physical structures. For instance in the case study on the Authie River (F), it was reported that most water mills nowadays are no longer in use, and are therefore neglected; sluices are not regularly opened, material is stuck in the mills and the river upstream is not cleaned, resulting in hydraulic clogging and increase of the river water level.
- According to the authors of the case study on the Baltic coast (S), the identification of modified hydromorphological (and biological) conditions in *coastal areas* was not as straightforward as in rivers. The case study on the Baltic coast (S) identified that along a coastline with a concrete quay, the shore and bottom in the immediate vicinity of the quay are changed, but the effects on the water and neighbouring shorelines are not equally apparent. A jetty connected to the quay affects primarily the environment in the area contained by the jetty. It is, instead, the activities associated with the quay (boat traffic, dredging) that affect the water environment over a larger geographical area.

Chapter 6

Ecological Status

6.1 Introduction

As shown in the identification and designation process for HMWB and AWB (Step 5 of Fig. 2.1, p. 12), the likelihood of failing to achieve Good Ecological Status (GES) (or an estimate of what GES may be, based on current knowledge) should be assessed (WFD Annex II 1.5). Such an assessment is based on the information gathered on significant changes on hydromorphology and an assessment of the ecological status. The final guidance document on HMWB and AWB also noted that it should be considered whether the risk of failing GES is due to hydromorphological changes and not other pressures such as toxic substances or other quality problems. Step 5 is part of the "risk assessment" process[1] to be completed by 22 December 2004 (WFD CIS Guidance Document No. 4 2003).

The WFD makes use of quality elements as principal components for the classification of the ecological status of water bodies. The quality elements for water bodies are listed in WFD Annex II (1) and include (*i*) biological elements, (*ii*) chemical and physicochemical elements supporting the biological elements and (*iii*) hydromorphological elements supporting the biological elements. The quality elements differ according to water categories.

This chapter presents the quality elements and assessment methods used to evaluate the likelihood of failing to achieve current ecological status in the HMWB case studies.

6.2 Biological Quality Elements

6.2.1 Quality Elements and Assessment Methods

According to the WFD Annex II (1), the quality elements differ according to the water categories. Therefore, in the following paragraphs the biological quality elements (and the methods for their assessment) used in the HMWB case studies are summarised according to the different water categories.

[1] The "risk assessment" is undertaken as part of the WFD Article 5 characterisation process and identifies the likelihood of water bodies to fail the environmental quality objectives set under WFD Article 4.

Rivers

The Water Framework Directive (WFD Annex V 1.1: Quality Elements for the Classification of Ecological Status) recommends using the following biological quality elements for classifying the ecological status of rivers:

- Composition and abundance of aquatic flora (macrophytes/phytobenthos, phytoplankton);
- Composition and abundance of benthic invertebrate fauna;
- Composition, abundance and age structure of fish fauna.

Thirteen out of twenty-eight of the HMWB case studies on rivers used *macrophytes/phytobenthos* as a quality element. In some cases, there was a lack of data on aquatic flora components as in the case of Nestos (GR), where a need for appropriate monitoring of the composition and abundance of macrophytes was indicated. In the cases of the rivers Ruhr and Dhünn (D), macrophytes were rejected as a suitable quality element because the case study authors considered they were of minor importance for mountainous rivers. In the case of Hagmolenbeek (NL), it was stated that it is only meaningful to monitor phytobenthos for ecological assessment in cases where nutrients are the limiting factor for the ecological quality of running waters.

Specific parameters used for the assessment of status based on macrophytes were:

- Composition and abundance;
- Estimate of cover;
- Species richness, as well as plant dispersal.

In several cases, data were assessed using expert criteria. In the case study on the Sarre River (F), analysis was based on the presence/absence of species indicating trophic quality, of species indicating eutrophy as well as lentic zone species.

In the case of Umealven (S), the bank vegetation of the river was compared with that of a neighbouring free-flowing river, which was used as a reference. Elements used included plant zonation, width of riparian zone, number of plant species per unit area, tree and shrub cover, and herb and dwarf shrub cover. For all elements, the % change against the natural state of comparison was calculated. A similar approach was used in the case study of the Tummel River (UK), where the number of species of edge flora and submerged/floating flora as well as % cover was compared between the study area and an area that was not impacted.

Only five out of twenty-eight of the HMWB case studies on rivers used *phytoplankton* as a quality element. In the case study on the Lagan River (UK), the need for detailed data of phytoplankton and phytobenthos in the context of the WFD was expressed. In the case of the Sarre River (F), the analysis of phytoplankton was based on the expression of eutrophication (chlorophyll a/Phaeo-pigments or expert opinion, taking algal blooms into account) and secondly the quality of the specific population (diatomic index of expert opinion).

All HMWB case studies on rivers used *fish* as a quality element. In general, no lack of fish data was reported apart from a few cases, such as the case studies on the rivers Lagan (UK) and Great Ouse (UK). It was reported that in the UK, no routine analysis is being applied to identify variations in fish density, biomass and species richness.

In some case studies, the use of simple parameters was made including the following:

- Composition/presence of species (frequently applied);
- Presence of typical rheophilic species (e.g. *Gobio gobio*, *Barbatula barbatulus*), species characteristic for slowly flowing streams, species characteristic for stagnant waters, presence of anadromous species (mostly in areas affected by hydropower);
- Abundance;
- Age structure.

Specific classification methods used for the assessment of biological status based on fish included:

- The use of seven biological criteria for fish (type-specific species, self-sustaining species, fish region, number of guilds,[2] guild composition, biomass and density, population age structure) and a definition of five levels of ecological integrity according to the normative classification of ecological status of the WFD; (A)
- Fish-index (Index of Biotic Integrity, IBI) based on species composition and richness, trophic composition, amount of fish and condition of fish population (the evaluation was combined with a comparison to the reference situation). The fish index (IBI) is a useful indicator for comparing historical and actual conditions; (B)
- The icthyofaunistical assessment procedure of the ARGE Elbe. This involves evaluation according to type inventory, abundance, age structure; (D)
- Comparison of historical and recent situation regarding the number of species, species similarity, and lack of species; (D)
- Comparison of the actual fish distribution with the hypothetic composition of natural fish fauna of the corresponding stream type. In this respect, six metrics were used: abundance, key species, anadromic and katadromic fish, key species barbel (of the reference condition), accompanying species, and share of key- and accompanying species. With regard to this method, the problem of lack of homogeneity of the fish databases was identified as well as the fact that there is no summary of the six metrics into a single status class of ecological quality; (D)
- Fish six-index level grid (from excellent to disturbed) taking into account species sensitivity to water and habitat quality. The deviation between the observed and the reference population was measured (the reference population was set by experts on the basis of fish integrity score figures); (F)
- In Spain, the Netherlands (Hagmolenbeek River), France (Sarre River) and the UK-EW (Tame River), expert criteria and knowledge were used to classify the biological quality elements of fish.

[2] A guild consists of species which take over the same function within the ecosystem, have similar life-forms with respect to habitat use, feeding habits, temperature preference etc.

Twenty-three out of twenty-eight of the HMWB case studies on rivers used *benthic invertebrate fauna* as a quality element. In the case study on the Nestos River (GR), no data on the composition and abundance of benthic invertebrate fauna were available. Only in a few other cases, lack of data was reported for the element of benthic invertebrate fauna. According to the authors of the UK-EW guideline report (Booker et al. 2002a), some catchments in the UK-EW had insufficient benthic invertebrate fauna monitoring sites to provide robust comparisons with equivalent sites of unmodified rivers and streams. It was noted however that since the UK has a dense riverine monitoring network relative to many other countries in Europe, it is likely that insufficiency of benthic invertebrate fauna monitoring sites may also be the case in other European countries.

Apart from assessing composition and abundance, the case studies used a mixture of methods and indicators which are summarised in Table 6.1 (for further explanations on the methods, the reader should refer to the individual case study reports).

In six of the twenty-eight case studies on rivers (rivers Bregenzerach, Danube and Wienfluss (A), rivers Elbe and Ruhr (D), and the Kennet River (UK)), *benthic invertebrate fauna* and *fish* were the only two biological elements used for classification. This was in accordance with the support material (guidance papers) agreed by the HMWB Working Group for conducting the case studies. The support mareial recommended that the assessment of ecological status should be based upon a minimum of two biological quality elements, e.g. invertebrates and macrophytes/phytobenthos or fish (depending on the water category and data availability). Regarding this issue, the final guidance document on HMWB and AWB suggested that the preliminary assessment of ecological status, to be completed by 2004, should be based on the most sensitive quality elements with respect to the existing physical alterations (WFD CIS Guidance Document No. 4 2003).

Lakes

The Water Framework Directive (WFD Annex V 1.1: Quality Elements for the Classification of Ecological Status) recommends using the following biological quality elements for classifying the ecological status of lakes:

- Composition, abundance and biomass of phytoplankton;
- Composition and abundance of other aquatic flora (macrophytes/phytobenthos);
- Composition and abundance of benthic invertebrate fauna;
- Composition, abundance and age structure of fish fauna.

Two of three case studies on lakes used *phytoplankton* as a quality element. In the case studies on the rivers Lozoya (E) and Nestos (GR), phytoplankton was assessed in the reservoirs of the case study areas. Similarly, in the case studies on the rivers Suldalslagen (NO) and the Emån (S), phytoplankton was assessed in the lakes of the case study areas. Parameters for assessment included concentration, representative genera, chlorophyll content (max. and annual mean), and expert criteria.

All HMWB case studies on lakes (3) used *macrophytes/phytobenthos* as a quality element. Parameters assessed included:

Table 6.1. Methods and indicators used for the assessment of benthic invertebrate fauna in rivers. *Source:* Compiled by the authors from the evaluation of the HMWB case studies

Method	Brief description/comments	Country
Analysis of the ecological integrity sensu ÖNORM M 6232 adapted to the WFD	Parameters: benthic invertebrate fauna species inventory, dominance structure (species, higher taxa), biomass, functional feeding groups, longitudinal zonation patterns, saprobic quality index	A
Saprobic index	The weakness of the saprobic index is that it is used to evaluate primarily pollution and was used in default of adequate indices. In Germany, it is still a problem to define ecological status adequately using benthic invertebrate fauna	A, D
Potamon type index	This index is limited to large potamal streams and cannot be used for all sections of a stream, i.e. not very suitable for upstream sections	D
Rheo index	Measure indicating flow conditions by ratio of specific species with an affinity to flow	D
AQEM	Multimetric method summarising impacts of stream morphology, saprobic status and other stressor-based metrics. The AQEM method needs quantitative data, which do not exist at present. However, the method can also be applied using existing data; this was done in the case studies Ruhr, Mulde and Dhünn and led to realistic results with the use of expert judgment. The method fits the requirements of the WFD for assessment with regard to reference conditions of water types	D
Comparison of present with reference condition	Metrics: species composition, functional group composition, saprobic index, diversity (Shannon-Weaver index)	D
IBGN (global biological index)	Based on taxonomic variety and fauna indicator groups	F
EBEOSWA (Ecologische BEOordeling van Stromende WAteren)	Combination of multivariate analysis of benthic invertebrate fauna data and expert judgment. It is based on abundance of species and distinguishes between general and specific species. It considers several environmental factors and provides classification of status in five classes. However, the method is based on streams in the Netherlands, many of which have been impacted by human activities (the Netherlands has one of the highest population densities in Europe). Thereby, the ecological quality results will possibly yield a more positive picture of the ecological level than assessment methods from other countries. The intercalibration exercise should make the differences between countries more clear	NL
RIVPACS system	Used to generate species lists for a particular site which are then compared against a large reference database of sites with similar physical attributes which are considered to be unimpacted. The difference between the site under investigation and the reference database sites gives a score of how impacted the site is. There is no scoring system which relates invertebrate communities to physical modifications, although one could be developed by linking RIVPACS to the River Habitat Survey system. In the absence of this, expert opinion, supported by reference species lists and knowledge of species habitat requirements should be used to assess the status with respect to benthic invertebrate fauna	UK
Presence of indicator taxa	Assessment based on the presence of indicator taxa such as Simuliidae, which require constant flow and "reasonable" water quality. Where water quality is adequate, their absence can be attributed to lack of water velocity	UK
Belgian biotic index	Presence/absence of pollution-sensitive benthic invertebrate fauna and diversity of species groups	B

Table 6.1. *Continued*

Method	Brief description/comments	Country
Dragonfly-based assessment of river/floodplain systems	Not strictly a benthic invertebrate fauna method. It is a classification scheme of five classes that uses the criteria: community structure (Odonate Habitat Index, autochthonous species) and taxonomic composition (number of sensitive autochthonous species, abundance of autochthonous species). It is linked to the habitat preferences of dragonflies (Odonata). The habitat types used follow a gradient of lateral connectivity with the main river channel ranging from permanent water bodies to isolated floodplain	A
Combination of number of taxa, number of individuals m^{-2}, Shannon diversity index	See Table 14.17 in Annex II	S
Expert criteria	–	E, F
BMWP (Biological Monitoring Working Party) index	–	UK
BMWP' (BMWP adapted for Spain)	–	E

- Number of species;
- Area of flooded *Carex* vegetation;
- Presence of frost-sensitive isoetids;
- Coverage.

Two of the three case studies on lakes used *benthic invertebrate fauna* as a quality element. Methods and parameters for status assessment in the case study on Lake Kemijärvi (SF) included:

- Number of typical taxons, number of typical taxonomical groups, number of taxons and presence of frost-sensitive species;
- Littoral benthic invertebrate fauna also considering pre-regulation data;
- Composition and abundance of species and taxon compared to data from non-regulated and slightly regulated other lakes.

In the Finnish case study, preliminary threshold values for the biological quality elements as well as Ecological Quality Ratios were calculated as shown in Table 14.19 (see Annex II).

All HMWB case studies on lakes (3) used *fish* as a quality element. Status of the fish community was assessed according to composition, presence of migratory species, abundance, and biomass.

Transitional Waters

The Water Framework Directive (WFD Annex V 1.1: Quality Elements for the Classification of Ecological Status) recommends using the following biological quality elements for classifying the ecological status of transitional waters:

- Composition, abundance and biomass of phytoplankton;
- Composition and abundance of other aquatic flora (macroalgae, angiosperms);
- Composition and abundance of benthic invertebrate fauna;
- Composition, abundance and age structure of fish fauna.

In the two HMWB case studies on transitional waters, fish, benthic invertebrate fauna and macrophytes were used as elements for the assessment of biological status. Phytoplankton was used in one case study.

The status of the *benthic invertebrate fauna* was assessed according to the number of characteristic species (comparison of current with reference) in the Haringvliet Estuary (NL) and according to expert judgment in the case on the Forth Estuary (UK).

The status of the *fish* community was assessed according to the number of species per ecological guild (comparison of current with reference) in the Haringvliet Estuary (NL) and according to expert judgment in the case on the Forth Estuary (UK).

In the case on the Haringvliet Estuary (NL), a new element was proposed as an additional quality element for the assessment of the biological status. This additional element is called *ecotopes*. Ecotopes are landscape ecological units, and their assessment is based on parameters of hydrodynamics, morphodynamics and land/water use (comparison of current with reference). Ecotopes are defined in such a way that the entire array of water systems and their surrounding areas of direct environmental influence are covered. As landscape ecological units, ecotopes function as ecologically meaningful units holding specific sets of habitats for specific species of plants and animals. By their nature (based on hydromorphology), ecotopes are directly related to measures in management and construction.

Coastal Waters

The Water Framework Directive (WFD Annex V 1.1: Quality Elements for the Classification of Ecological Status) recommends using the following biological quality elements for classifying the ecological status of coastal waters:

- Composition, abundance and biomass of phytoplankton;
- Composition and abundance of other aquatic flora (macroalgae, angiosperms);
- Composition and abundance of benthic invertebrate fauna.

The only HMWB case study on coastal waters (Baltic coast, S) used mainly *benthic invertebrate fauna* as a quality element as well as some data on fish, even though fish are not recommended by the WFD as a biological quality element for classifying the ecological status of coastal waters. The case study authors expressed the need for a complementary monitoring programme and consistent assessment methods for the archipelago and near-shore shallow coastal areas, since the national monitoring covered mostly the open sea.

6.2.2
Impacts of Specified Uses and Physical Alterations

In the following paragraphs, impacts upon the biological elements have been summarised according to the main specified uses in the HMWB case studies. Impacts are also presented and described separately for some types of physical alterations that are relevant to more than one specified use. For instance, weirs are physical alterations associated with both hydropower and navigation.

Linking hydromorphological changes to biological impacts was rarely straightforward in the HMWB case studies. Assessments of impacts upon biology were mainly based on expert judgment using the biological assessment methods of Sect. 6.2.1 of this book. Specific approaches for linking hydromorphological changes to biological impacts and thus separating physical impacts from water quality impacts are mentioned in Sect. 6.2.3.

Specified Use: Hydropower

In the HMWB case studies, impacts of hydropower development on *fish* resulted from changes in water quality, water temperature and availability of habitat especially for spawning. Additional issues for migratory species were the restriction of access by physical obstructions and reduced flows, loss of directional cues due to flow regulation, predation of delayed or disorientated fish, and injury/mortality in unscreened turbines.

From the evaluation of the HMWB case studies, the following can be concluded concerning fish:

- Changes in habitat favour the introduction of new, better-adapted species that unbalance the local fauna;
- Hydropower plants are obstacles to migration reducing the amount of fish upstream of dams and suitable biotopes for species such as salmon, sea trout and migrating whitefish;
- The fish habitat can be affected by the accumulation of sand, which clogs interstices between cobbles and boulders as well as the increase of moss growth related to low flow and absence of large floods.

Among the impacts of hydropower on *benthic invertebrate fauna*, the following were reported in the HMWB case studies:

- The abundance of benthic invertebrate fauna is reduced due to decreased water flow and decreased water cover (reduced production area);
- The downstream transport of benthic invertebrate fauna is interrupted due to dam barriers;
- In rapids with reduced flow, abundance of collectors, grazers and predators are lower than in free-flowing sections;
- Downstream of dams, the type of substrate changes significantly and the amount of silt increases. The benthic invertebrate fauna can disappear and be replaced by very poor communities made up of organisms adapted to living in silt;
- Dominance of insensitive taxa (chironomids) may be noted in residual water reaches at the expense of more sensitive taxa such as Plecoptera (reported in the Austrian case study on the Bregenzerach River). Long-living and oxybiont benthic inverte-

brate fauna are periodically lacking, while insensitive, phytorheophile organisms dominate;

- Caddisflies and filter feeding taxa are strongly reduced due to the lack of drifting food and the clogging of drifts due to silting as a result of regulation. The increased moss cover, after regulation, affects the bottom structure, intra-gravel and bottom hydraulics leading to chirono-mids being the dominant fauna;
- Some sensitive species of benthic invertebrate fauna may disappear in the upper littoral of northern regulated lakes (as in the Finnish case study) due to ice and freezing of sediment.

The creation of impounded water bodies permits the development of *phytoplankton* communities. In the case study of Spain (Lozoya River), however, the limited capacity of the reservoirs (low retention time) limited the development of the phytoplankton community.

In the Greek case study (Nestos River), little diversity was observed in the reservoir upstream of the dam, which was attributed to the lake instability. The large inflows and outflows and the great fluctuations in water level on a daily basis cause major rearrangements and extinction of the developing ecosystem.

Among the impacts of hydropower on macrophytes, the following were reported in the HMWB case studies:

- Aquatic vegetation becomes notably scarce, with only a few species characteristic of lentic waters remaining;
- Strong, sudden fluctuations in flow downstream of dams reduce or eliminate natural vegetation. Similarly, the reservoir shores have scarce vegetation;
- River margins along regulated stretches lack the distinctive zonation of free-flowing rivers;
- Moss/liverworts vegetation *(Fontinalis* sp., *Marsupella aquatica)* may increase after regulation as in the case study on the Suldalslagen River (NO), where the increase was attributed to strongly reduced frequency of large floods, increased winter low flows and reduced ice cover;
- The downstream transport of plant diaspores is interrupted due to dam barriers. The function of the river and riparian zone as pathway for plant dispersal is lost and as a consequence, riparian flora is more similar among sites within an impoundment than among sites between impoundments;
- In northern lakes (as in the Finnish case study), several common species of macrophytes may disappear or decrease after regulation, such as species sensitive to increased ice and erosion. In the Finnish case study, the area of littoral vegetation (*Carex* species) also decreased dramatically.

Specified Use: Navigation

Dredging (for the purpose of navigation) decreased the area of shallow water in rivers that are spawning and nursery grounds for *fish*. The same was reported for coastal waters, where important areas for fish reproduction were disturbed by increased turbidity and sedimentation. In the Elbe River (D), living conditions for most *benthic invertebrate fauna* communities were unfavorable due to increased movement of riverbed material. Waves, passing boats as well as dredging for the purpose of navigation also influenced the ecology of the benthic invertebrate fauna in rivers, transitional waters and coastal waters.

In the case study on the Baltic coast (S), it was reported that the substrate for *macrophytes* was damaged by passing boats.

Specified Use: Flood Protection

According to the authors of the case study on the Lozoya River (E), linear physical alterations, such as canals and flood-works have a smaller impact on biological quality elements than structures that interrupt the longitudinal continuity. This is because they change only part of the habitat and are considerably less extensive. The organisms basically affected are the benthic invertebrate fauna and the aquatic vegetation, which respond with a decrease in the biodiversity by simplification of the spatial heterogeneity. On the other hand, the interruption of longitudinal continuity is related to the biological isolation of separate sectors leading to the presence of exclusive species.

Fish were affected by the lack of connectivity, canalisation and the reduction of spawning/refuge habitat. The low flow conditions and the loss of regular flooding events affected the spawning success of phytophiles, i.e. fish which spawn on plants. The presence of physical barriers (e.g. weirs, locks), was a major factor affecting migratory fish species. This either excluded species, or at least reduced their numbers markedly. The predominately vertical river banks did not allow a successful colonisation by marginal vegetation, which could be used as refuge habitats for fish larvae as well as a good spawning habitat for phytophilic fish species (Great Ouse River, UK). Similarly, in the Lagan River (UK), a habitat for coarse fish was missing due to high winter flows and lack of marginal refuges in high flows.

Flood defence modifications affected the seasonal variation of the water level and as a result affected the spawning success of species which depend on flood events (pike, perch). The lack of inundated floodplain affected species that favour low flow and submerged vegetation as spawning habitat (rudd, bream). Straightening reduced marginal vegetation and areas of low flow conditions like oxbows, which in turn reduced coarse fish larvae and increased young-of-the-year mortality (Tames River, UK).

In the case of the Haringvliet Estuary (NL), which was affected by a dam for flood protection, typical estuarine species were almost absent. Free fish migration was hampered by the dam.

In the case study on the Haringvliet Estuary (NL), all *benthic invertebrate fauna* species characteristic of brackish and slightly brackish ecotopes were absent due to the management of the sluices and the dam. In general, only a few ecotopes characteristic of estuarine ecosystems existed due to the lack of salt water intrusion.

High peak discharges were related to the absence of *macrophytes*. In the Dutch lake case study on the Veluwerandmeren, it was reported that no typical marsh vegetation grew on the banks due to bank fixation, the unnatural steepness of the banks and the unnatural water level management.

Specified Use: Urbanisation

Urbanisation in the Sankey catchment (UK) has led to the loss of marginal habitat for *fish* spawning partly due to a lack of longitudinal habitat diversity, impoundment and lack of floodplain channels. In the urbanised areas of the Lagan River (UK), habitats for salmonids were reduced due to a highly silted bed and the extraction of gravel from the river. Entrapment into water intakes for industry was also reported as a cause of fish mortality in urban industrialised areas.

Land-take in connection with urbanisation and industry led to loss of *macrophytes* and significant habitats for biodiversity purposes such as salt marsh in areas of coastal and transitional waters.

Physical Alteration: Reservoirs

Reservoirs and impoundments are physical alterations that can be associated with hydropower as well as navigation, flood protection and indirectly with recreation.

The *fish fauna* downstream of reservoirs was poor due to large fluctuations in water levels and the low minimum flow, as observed in the Spanish case study. The presence of impoundments and reservoirs also reduces spawning grounds due to siltation and lack of gravel banks and leads to the replacement of river species with lake species.

Downstream of a reservoir, a stable biocoenosis of *benthic invertebrate fauna* could not develop due to the artificial flow regime. In the reservoirs, there was only benthic invertebrate fauna typical to fine sediment substrates. Big losses of benthic invertebrate fauna also occurred due to large water level fluctuations.

Physical Alteration: Weirs

The presence of weirs and the consequent disruption of river continuum impacted *fish* species that migrate large distances such as salmon and lake trout.

Regarding benthic invertebrate fauna, in the case of the Lahn River (D), it was concluded that weirs have contributed to a loss of shredders due to hydraulic pressure and an increase of bioturbators due to high solid loads. A significant reduction of the number of species and of filter feeders above the weirs was noted.

In the case of the Authie River (F), the silt substrate above the weirs limited the development of predator benthic invertebrate fauna living on rocky substrate. Sensitive taxa that are characteristic of salmonid rivers were rare, while burrowers and filterers proliferated.

Physical Alteration: Drainage

In the case study of the Great Ouse (UK), it was noted that drainage has contributed to a reduction in the quantity of *fish*, mainly because it left small drains and ditches, which used to contain abundant pike, eels and perch, completely dry in the summer.

Drainage was also reported to lead to a decrease of wet habitats, such as reed grassland and wet meadows.

6.2.3 Other Impacts

Impacts from Pollution

In at least sixteen HMWB case studies, biological elements were affected by pollution originating mostly from untreated sewage, agricultural run-off as well as acidification. In several cases, fish were affected by low water quality and low oxygen levels. Pollution was also often responsible for the absence of aquatic plants. Where dams existed, eutrophication led to a strong development of seaweed in lentic zones. The growth of blanket weed in shallow sections close to wastewater discharge points was also reported. Regarding the phytoplankton community, the phenomenon of "algal bloom" was reported in the case of the Forth Estuary (UK) and attributed to nutrients and sewage treatment. In hot summer months, such algal bloom may cause changes in the benthic invertebrate fauna community from a multifunctional to a collector-gatherer and filter-feeder community.

Other Impacts

Fish stocking seemed to constitute a pressure affecting fish population in the case of the Suldalslagen River (NO), Great Ouse (UK) and the Dee River (UK).

How Were Physical Alteration Impacts Separated from Other Impacts?

According to the Spanish case study, it is considered easy to distinguish between the response of organisms to the effects of physical alterations and the effects of other pressures. In general, the changes observed in response to physical alterations are quite evident, as they have to do with changes in the habitat. The biological changes produced by other alterations such as pollution or depredation have a greater reversibility. However, this was not the view of many other Member States such as the UK-EW, who stated that it was very difficult to separate biological impact resulting from different types of use and pressure. This is particularly the case when different pressures affect the same parameters (e.g. the non-physical modification impact of water abstraction versus the impact of a physical barrier on flow; or sediment clogging from point or diffuse sources that may not involve physical modifications versus the physical impacts of flow regulation caused by reservoirs on sediment clogging).

The UK-EW guideline report suggested that a new method should be developed linking the River Invertebrate Prediction and Classification System (RIVPACS) with the River Habitat Survey Methodology (RHS) used in the UK. RIVPACS is a model that contains an extensive database of reference sites that are compared to study sites on the basis of hydromorphological and physicochemical parameters to predict what invertebrates should be present. RHS is a system that calculates habitat quality and degree of modification on the basis of a standard survey. This new method would be qualified by information and expert opinion on reference species lists and knowledge of species habitat requirements.

In the case study of the Emån River (S), it was proposed to use different benthic invertebrate fauna indexes, such as the *acidification index* and *Danish fauna index* for acidification and organic pollution respectively, in order to separate impacts of physical and chemical influences.

In the case study on the Baltic coast (S), eutrophication was the major cause of effects on flora and fauna in the area, but it was difficult to distinguish the effects of eutrophication and toxic substances from effects of physical alterations. According to the case study authors, in the open and complex archipelago ecosystem, a range of anthropogenic pressures interacts with constant natural development (succession), making separation of specific effects of certain activities impossible.

In the German case studies, the value of a new methodology called AQEM was pointed out (EU research project "Development and testing of an integrated assessment system for the ecological quality of streams and rivers throughout Europe using benthic macroinvertebrates"). This method allows the establishment of relationships to specific pressures and impacts that cause changes to macroinvertebrate communities. According to the authors of the German case studies, an appropriately modified version of AQEM could be suitable for the assessment of the Good Ecological Potential of HMWB with regard to macroinvertebrates. In this case, the reference conditions need to be redefined and single "metrics" modified.

6.3 Physicochemical Quality Elements

6.3.1 Quality Elements and Assessment Methods

Regarding the use of physicochemical elements supporting the biological assessment, approaches in the case studies varied according to the availability of data. In general, the case studies made use of the elements required by the WFD for the corresponding water categories (as required by WFD Annex V 1.1.5). The most extensively used elements in rivers were thermal conditions, oxygenation conditions, nutrients, and acidification. Transparency was assessed in the three lake systems as well as in cases where reservoirs existed. In a few cases, specific and non-specific synthetic pollutants were assessed. In the case studies, use was made of the databases of national monitoring systems and water quality surveys. In the case of the Nestos River (GR), it was recommended that existing data need to be complemented for the purposes of the WFD. In the case on the Baltic coast (S), the need was expressed for a complementary monitoring programme for the archipelago and near-shore shallow coastal areas, since present national monitoring predominantly covers the open sea.

Data were most often assessed using expert criteria as well as existing classification methods such as the chemical window of the General Quality Assessment Method (GQA) in the UK-EW, which also uses five quality classes. In a German case study (Seefelder Aach), an existing chemical classification system (LAWA 1998) of seven quality classes was transformed into five classes to be in accordance with the WFD.

6.3.2 Impacts of Specified Uses and Physical Alterations

The following paragraphs summarise the impacts of specified uses and physical alterations on the physicochemical elements examined in the HMWB case studies.

Specified Use: Hydropower

Various effects on water quality likely to affect living organisms have been associated with hydropower operation. The effects centre around the fact that turbine intakes are often near the bottom of the reservoir, where conditions of aeration and temperature may differ from those at the surface. Such effects are particularly marked when the reservoir is stratified; the water below the surface mixed layer is then relatively cold and dense with a low concentration of dissolved oxygen. The effects of the physical alterations are reflected in the temperature, which is lower than normal in the summer, and possibly in the lower dissolved oxygen in the water drained from the reservoirs during the most eutrophic episodes.

In the Nestos River (GR), the dam operation caused great losses of water, which affected dissolved oxygen conditions and the conditions of H_2S emissions. H_2S is formed in autumn due to the large water abstractions and low inflow in the summer.

In Lake Kemijärvi (SF), water level regulation and high discharge from reservoirs broke winter stratification and impacted oxygen conditions.

Specified Use: Flood Protection

In the case of the Haringvliet Estuary (NL), the dam constructed for flood protection purposes led to the development of a highly contaminated sediment layer and the disappearance of the estuarine salinity gradient.

In the case of the Great Ouse River (UK), elevated levels of dissolved oxygen occurred downstream of flood protection structures together with a lack of warm backwaters because of bank uniformity.

Specified Use: Navigation

In the case of coastal waters, which are under a mixture of uses such as navigation, coastal protection, urbanisation and recreation (see the case on the Baltic coast, S), the water became turbid due to shore and bottom erosion. The presence of jetties reduced water circulation, which decreased oxygen levels.

In rivers, impoundments for navigation extended water residence time, and during low summer flows the associated phytoplankton blooms resulted in overnight falls in oxygen concentrations.

Impacts from navigation were also reported for transitional waters. In specific, sea temperature increased adjacent to shipping channels and power station facilities. Dredging and shipping additionally recirculated heavy metals deposited in sediments.

Physical Alteration: Reservoirs

The main impacts of reservoirs on physicochemical elements reported in the case studies were modified temperature regime/stratification of water, night-day and seasonal changes in pH, and change of oxygenation conditions in stratification season.

According to the Spanish case study, reservoirs act as both accumulators and dissipaters of heat, and therefore, along the course of the river there is a homogenisation of the water temperatures and a reduction of the annual and interannual thermal fluctuations. In the summer and fall, the water discharged from the reservoir has a temperature lower than normal due to the thermal stratification, while in winter the water discharged is more temperate.

Physical Alteration: Weirs

According to the case study on the Lahn River (D), extended backwaters above weirs led to increasing loads of nutrients towards weirs. This effect depended on wastewater inflow and failing catabolism in the standing and often stratified backwaters.

6.3.3 Other Impacts

In numerous case studies, the presence of organic and nutrient pollution was reported due to wastewater discharges from industries, human settlements, and diffuse input from agricultural land, livestock and contaminated land leakage. In some cases, industry was also a source of heavy metal pollution while mining activities were reported as a source of acidification.

Low oxygen levels were reported in connection with organic pollution due to sewage water, even downstream of fish farms. Eutrophication degraded water quality in rivers and lakes as well as in coastal waters.

6.4 Hydromorphological Quality Elements

According to the WFD Annex V, hydromorphological quality elements should be used as part of the definition of *high* ecological status. Some of the HMWB case studies have used hydromorphological quality elements as "parameters indicative of biological quality elements" to define current ecological status. Within this context, the purpose of this section is to report on the use of hydromorphological elements explicitly mentioned and used for the definition of current ecological status in the case studies. The overall parameters and methods used with regard to the assessment of the hydrological and morphological status of the water bodies have already been presented in detail in Chap. 5. Therefore, they will not be repeated here.

In the case study on the Emån River (S), the hydromorphological criteria from the System Aqua of the Swedish Environment Protection Agency were used: assessment of permanent modification, modification of flow, fragmentation, and land use (see Table 14.17 in Annex II).

In the Umealven River (S), spring flood and summer low levels as well as flow diversions were used as criteria of hydromorphology in the assessment table of current ecological status (see Table 14.18 in Annex II).

The case studies on the rivers Hagmolenbeek (NL) (see Table 14.16 in Annex II), Suldalslagen (NO), Beiarn (NO), and Dee (UK) made use of all hydromorphological elements of the WFD for rivers, i.e. hydrological regime, river continuity and morphological conditions.

According to the authors of the case study on the Dee River (UK), "river continuity" is a composite concept. It affects several factors related to ecological status:

a The ability of fish to gain access to the water body;
b Their ability to move upstream from its upper limit;
c Their ability to move downstream from its lower limit;
d The supply of sediment to the stretch of river in question;
e Trapping of sediment and sediment starvation of the reach downstream.

Thus, a single structure can have multiple and differential effects. For example, an impoundment provided with a fish pass at the upstream end of the water body in question may afford acceptable quality in respect of *(b)* but be unacceptable in terms of *(d)*, whilst a "farthest-upstream" impoundment without a fish pass will influence factor *(a)* for water bodies upstream. The latter consideration also has significant implications for the definition of water bodies. According to the authors of the case study on the Dee River (UK), in terms of all other criteria, it is necessary to consider impacts only on those sections of streams and rivers that lie downstream of artificial structures. The "upstream continuity" factor requires consideration of the whole river network upstream of any structure, although there is potential for its demotion from high ecological status on the grounds of only one biological quality factor. This quality factor is the distortion of the composition of the fish population, but only if proof can be given that the fish impeded by the structure would use the river upstream under natural conditions (natural accessibility)[3]. According to the case study authors, if the upstream section is not physically modified but fails to reach GES, it should not be considered for HMWB designation but for a derogation of "less stringent environmental objectives" (River Dee, UK). The authors of the case study also noted that in terms of further research needs for such cases, there is a need to apply ecological standards and species access criteria.

6.5 Assessment of Current Ecological Status

6.5.1 Overview of Methods

Different combinations of biological, physicochemical and hydromorphological elements were used in the case studies for the purpose of ecological status classification:

- In ten case studies (rivers Bregenzerach, Danube and Wienfluss (A), rivers Elbe, Lahn and Ruhr (D), Hagmolenbeek River (NL), Umealven River (S), Kennet River (UK), and Sarre River (F)), only biological elements were used for the classification of ecological status. In six of these ten (rivers Bregenzerach, Danube and Wienfluss (A), rivers Elbe and Ruhr (D), and Kennet (UK)), benthic invertebrate fauna and fish were the elements used for classification;
- In six cases (Lozoya River (E), Authie River (F), Haringvliet Estuary (NL), Beiarn River (NO), Sankey Brook (UK), and Lagan River (UK)), a combination of biological and physicochemical elements was used to classify current ecological status;

[3] Natural accessibility involves the use of individual tributaries by migratory species and is governed by their ability to negotiate steep sections of the channel (tributary specific).

- In two cases (Emån River (S) and Great Ouse River (UK)), biological and hydromorphological elements were used;
- All ecological quality elements (biological, physicochemical, and hydromorphological) were used in three case studies (Suldalslagen River (NO), Forth Estuary (UK), Rhone River (F)).

More elaborate methods for the assessment of the current ecological status included the following:

- In the Belgian case study (Dender River), a rating system was applied for each of the identified water bodies of the river. The evaluation of ecological status was based on different indicators including the Belgian biotic index, fish index, PBI index, structural quality index as well as assessment methods on underwater soils, and aquatic and riparian plants. Each parameter can range between the value of 0–5. The sum of values of the individual parameters is classified into five classes of ecological quality reflecting the classes in the WFD.
- In the case study on the Emån River (S), *ecological status classification systems* were applied for lakes and rivers including several quality elements (see Table 14.17 in Annex II). Each of the quality elements was classified into one of the five classes. For each of the water bodies and reference sites, the mean values of the applied biological and hydromorphological quality elements were calculated. "Good status" was defined as being equal to the mean value of the reference site, while "high status" was defined as being 10% higher than the value for good status.
- In the case study on the Umealven River (S), the five classes of ecological status according to the WFD were interpreted as *quantifiable classes for a number of ecosystems, community, population and hydromorphological parameters* (see Table 14.18 in Annex II). For this purpose, parameters of central importance for the ecological integrity of rivers were chosen as indicators of natural processes and of the ability of the system to retain viable populations of naturally occurring species. All the criteria listed in the table must be met for each class; otherwise, the water body is assigned to a lower class.[4]
- The general procedure proposed by the UK-EW for the assessment of ecological status entails *a comparison with the reference condition*. The five qualitative classes of the WFD have been quantified based on a defined deviation from the reference condition (see Table 6.2). The deviation should be assessed for each quality element and sub element each time. A table was provided with the proposed quality elements and the potential scoring elements (Table 6.3).
- In the case studies on the Forth Estuary and the Dee River (UK), use was made of *physical quality indicators as surrogates for biological impact*. In the case study on the Forth Estuary, threshold levels were defined for the extent of the biologically significant physical alterations at which Good Ecological Status is not achieved (defi-

[4] With regard to the proposed criteria where riparian vegetation played the most important role, it was noted that riparian plants may be good indicators for hydropower but may poorly predict ecological status of other elements such as fish and benthic invertebrate fauna. The risk therein is that future efforts to increase ecological status may focus on the chosen criteria alone. However, if process-related criteria are used, this risk may be minimised.

Table 6.2. Deviation of ecological status from reference in the UK-EW. *Source:* UK-EW guideline report (Booker et al. 2002a)

Ecological status	Level of deviation from reference
High	0– 5%
Good	6–25%
Moderate	26–50%
Poor	51–75%
Bad	>75%

Table 6.3. Quality elements for the assessment of ecological status in the UK-EW. *Source:* UK-EW guideline report (Booker et al. 2002a)

Quality elements	Quantitative elements
Habitats	Scored deviation in the following categories: Surface flow type Channel substrate Channel features Bank features Land use Trees Special features Overall HQA Supported by data on modification types for left and right bank (from RHS/FDMS/other data)
Plants	Taxonomic composition and balance between species Supported by bank vegetation structure and channel vegetation from RHS surveys
Fish	Taxonomic composition and abundance and biomass (where data available)
Macroinvertebrates	Taxonomic composition and/or indicator taxa

nition based on expert knowledge). For instance, loss of up to 30% of the intertidal area and modification of up to 50% of the shoreline is compatible with Good Ecological Status (GES). In the case of the Dee River (UK), the physical surrogates comprised of hydromorphological and physicochemical elements. This method assumed that ecological status is at least good unless there is evidence to the contrary. Both the existence of direct evidence for the distortion of the biological elements as well as the risk of distortion of biological elements based on physical surrogate elements was considered. The authors of the latter case study commented that it is debatable whether biological data could offer clear indications of physical problems in a catchment like the Dee where the performance of biota is also affected by water quality problems (acidification). Moreover, in view of the time-consuming nature of biological sampling and data analysis, the case study authors considered that an ability to make quick physical assessments of ecological status throughout catchments will always be of use in targeting biological effort.

In the case study on Lake Kemijärvi (SF), the assessment of ecological status was also based on indirect criteria based on physical parameters (REGCEL water level analysis, see Sect. 5.4 of this book for more details).

- Similarly, in the case study on the Dalälven River (S), the definition of criteria to describe ecological status for fish did not incorporate any actual fish variables (Table 14.20 in Annex II). In order to use actual fish variables, more work would be needed in connection with class limits etc. Moreover, to more correctly estimate effects of hydropower and regulation on fish within the Dalälven River, more field investigations would be required in the vicinity of hydropower stations and in reservoirs.

  Expert judgment and information from literature formed the basis for defining relevant variables concerning the impact on fish using different scales or target units. According to the case study authors, the strength of the proposed system is that it allows a choice of approach relevant for the scale and level of detail required for evaluating the status, from river catchment to a smaller scale such as a river reach or lake. The main weakness is that biological elements (fish) were not tested directly against the proposed criteria.
- In at least eight case studies, expert judgment was used to assess current ecological status.

The final guidance document on HMWB and AWB suggests that in order to assess the likelihood of failing to achieve Good Ecological Status (GES), the ecological impacts of physical alterations on the water bodies in question should be estimated (WFD CIS Guidance Document No. 4 2003). The effort expended in the assessments should be proportionate (i.e. a tiered assessment approach should be used). For water bodies that are likely not to achieve GES (e.g. water bodies that have changed category due to physical alterations), effort expended estimating GES should be limited and conclusions of non-achievement of GES should be rapidly reached. In these cases, more effort can be expended in assessing Good Ecological Potential (GEP) early and the risk of not achieving it could be investigated. Likewise, through risk screening, a conclusion on excluding those water bodies that are clearly going to reach GES from the HMWB or AWB identification and designation process should be reached early and with minimal effort (WFD CIS Guidance Document No. 4 2003).

6.5.2 Selection of Reference Conditions

From the evaluation of the HMWB case studies, it was made obvious that the selection of the reference condition is a crucial step in the HMWB provisional identification process; it is used as the basis to determine the deviation of the current status from Good Ecological Status. The guidance proposed by the Common Implementation Strategy working groups 2.3 and 2.4 *on "Reference conditions for inland surface waters"* (WFD CIS Guidance Document No. 10 2003) and *"Typology, classification and reference conditions for transitional, coastal waters"* (WFD CIS Guidance Document No. 5 2003) should help define reference conditions for non-physically modified water bodies.

In some HMWB case studies, it was explicitly mentioned that it was difficult or impossible to define and select appropriate reference conditions mainly due to the lack of data and standards:

- In the case of the Nestos River (GR), it was reported that monitoring of comparable water bodies needs to be conducted in order to find an appropriate reference system. Especially regarding the reservoirs of the case study area, their geometry and operation hardly simulate a natural system. Moreover, their operational time has been short; therefore, the system has not yet reached an equilibrium.
- In the Norwegian case studies (River Beiarn and River Suldalslagen), it was stated that there is a need for data and knowledge from unregulated catchments. The Norwegian landscape is very varied in topography, geology, and land use, and representative reference sites are not easily found. Therefore, assessments need to rely on a fair amount of expert judgment.
- In the rivers Tame and Dee (UK), there was lack of historical data and reference sites. In the case of the Forth Estuary (UK), it was also impossible to identify another physically similar estuary that is undisturbed and where data are available. In these cases, a high degree of expert opinion is needed to define reference conditions.

In nine HMWB case studies, comparison with a reference condition was made, which was theoretical, historical or existing, in order to assess current ecological status. Approaches followed are described in more detail in the following paragraphs according to the different water categories.

Rivers

In the HMWB case studies dealing with river systems, the search for adequate existing reference sites concentrated along the same river or alternatively, in another river/floodplain system in the same eco-region. The indirect definition of a reference community was based on historical abiotic and faunal data or even existing literature for the investigated river type. A combination of historical data, river typology and comparison with similar rivers to establish reference conditions was encountered.

In several cases, the use of existing reference sites was made as in the case study on the Nestos River (GR). The authors of the latter suggested that regarding the water bodies downstream of the dams, reference conditions could be determined by comparison with the water body upstream of the dams. Regarding the Nestos delta region, natural conditions could be determined by comparison with another delta. Other existing lakes and rivers in the catchment were also used as reference in the case study on the Emån River (S). In the latter case study, it was recommended that reference areas for rivers should be rivers of the same stream order as the one under investigation. In the case study on the Wienfluss (A), a tributary system of the same river type was selected as a reference, due to the lack of a section of high ecological status as a reference condition. In this context, the assessment for each indicator group used in the study (fish and macroinvertebrates) was performed by using a type-specific approach.

According to the authors of the case study on Umealven (S), rivers with high ecological status still exist in Sweden or can be inferred with reasonable certainty. In the case of the Umealven River, an existing neighbouring river was used as a reference. The case study authors noted that this situation was of great advantage, allowing the quantification of the deviation of the Umealven River from natural conditions in the absence of hydroelectric exploitation.

In the UK-EW guideline report (Booker et al. 2002a), a specific methodology for the selection of reference sites was outlined. Appropriate reference sites should be selected by matching several site-specific "map-based" attributes from the River Habitat Survey (RHS) reference network using the procedure described in Box 6.1.

Lakes

From the three case studies that focused on lakes (Lake Kemijärvi (SF), Veluwerandmeren and Loosdrecht (NL)), only the two Dutch case studies discussed the issue of reference condition selection extensively. Here a summary of the relevant discussions is presented:

- In the case study on Lake Loosdrecht (NL), the different phases of the history of the lake were described. The historical phase of a strongly nutrient-limited, oligotrophic lake was chosen as a reference condition because:
  - It was the first phase after the creation of a shallow freshwater lake in its present form;
  - It was a phase with little human impact. The mesotrophic condition, which followed the oligotrophic one, was the result of eutrophication of the inlet water and drinking water extraction; therefore, it was a disturbed condition.
- In the case of the Veluwerandmeren (NL), a theoretical reference condition was chosen corresponding to a shallow, clear mesotrophic lake with natural water levels and natural banks and marshes (based on literature). According to the authors of this case study, selection of a historical reference condition was difficult because a refer-

Box 6.1. Selection of Reference Sites in the UK-EW

1. Select the potentially 'best reference' sites by comparing each of the River Habitat Survey (RHS) sites in each test section to the nearest 100 sites from the dynamic RHS classification system, i.e. in PCA space, from the main RHS reference data set using the four map-derived parameters: height of source, distance from source to site, height of survey site, and slope of stream bed at survey site (Jeffers 1998). For sections under consideration with more than one RHS site, remove duplicate data.
2. Match RHS site in section under consideration with semi-natural reference sites only (Habitat Modification Index < 3) for each test section.
3. Reduce potential reference sites to only three RHS sites by removal of those not similar in the following order:
   - Deep and surface geology;
   - Distance location (i.e. another catchment or Environment Agency region); and,
   - Different physical characteristics (width, RHS flow types, discharge category).

If an abundance of sites remain, further reduce numbers by referencing to maps of predominant adjacent land-uses associated with higher population densities (e.g. urban, dominating channel features).

Source: UK-EW guideline report (Booker et al. 2002a).

ence could be an estuary or a freshwater lake according to the history of the creation of the Veluwerandmeren. A historical reference estuary was considered inappropriate, because there was a change of water category to a freshwater lake and there were no plans to return to the estuary condition (by removing the dam) (contrary to the discussion on the Haringvliet Estuary, see below).

A (historical) reference as freshwater lake was not appropriate either, since the ecological condition had changed continuously since the lake's creation; the mesotrophic condition after the lake's creation was also disturbed due to fortified banks and unnatural water level management. Therefore, a theoretical reference condition corresponding to a shallow, clear mesotrophic lake with natural water levels and natural banks and marshes was chosen, because it represented the condition after the creation of the Veluwerandmeren in its present state.

The overall conclusions of the Dutch case studies on this issue are summarised:

- When former hydromorphological changes have led to a change of water category (e.g. change from estuary to freshwater lake due to the construction of a dam), the reference condition should be based on the new water category (undisturbed freshwater lake). This results in the reference condition not being based on an undisturbed historical situation but on a spatial or theoretical based reference.

  However, if the present policy aims to "remove" the hydromorphological changes and to restore the former water category, the reference condition should be based on the former water category and can be historically based (e.g. the reference condition in the case study on the Haringvliet Estuary (NL) was an estuary, as it had already been decided that the Haringvliet dam would be opened in the future).
- Many lakes have different phases in history with a different level of hydromorphological impact and a different ecological condition of even a different water category. Therefore, it is not always very clear which phase represents the undisturbed reference condition. The selection of a certain phase as a reference condition has consequences for the designation as natural or HMWB. This was exemplified in the case study on Lake Loosdrecht:
  - Selection of the oligotrophic condition of ±1920 as a reference condition would result in the lake not reaching the GES in its present condition and in being designated as HMWB;
  - Selection of the mesotrophic condition of the period 1930–1955 as a reference condition would lead to a situation where the lake is near the GES before 2015. The mesotrophic condition was the result of eutrophication of the inlet water and drinking water extraction; therefore, this condition was not undisturbed and was not selected as a reference.
- The ecological functioning of shallow freshwater lakes is the product of their hydromorphological environment. Especially in an undisturbed state (when the human impact such as eutrophication has been removed) the hydrology, morphology and ecology of each lake is very specific. For example, the oligotrophic reference condition of Lake Loosdrecht was the result of the creation of the lakes at that specific location as a result of human influence. A natural lake would never have been

formed there. Therefore, it was difficult to define a reference condition that could apply to a number of lakes. The risk is that either the reference condition may be too detailed and differ too much from the "real" undisturbed condition of a lake or the description of the reference may be too general and not suitable for the ecological assessment of the lake.

Transitional Waters

As mentioned above, in the case of the Haringvliet Estuary (NL), the reference condition was an estuarine situation (historical situation without flood defences), since it had already been decided that the Haringvliet dam would be opened in the future.

6.5.3
Results on the Definition of Current Ecological Status

Out of the thirty-four HMWB case studies, twenty-nine case studies explicitly mentioned the final result of the assessment of current ecological status as required by the WFD. In five case studies, although the individual quality elements were assessed, no explicit reference was made to the overall value of current ecological status, i.e. high, good or lower than good. The results of the assessment of current ecological status as well as of the individual quality elements (biological, physicochemical and hydromorphological), as far as available and mentioned in the case studies, can be seen in more detail in Table 14.14 (see Annex II).

Ecological status was mainly defined for each of the identified water bodies except in the case study on the Lozoya River (E) where it was defined for water body groups. In summary, either some or all water bodies in each HMWB case study failed to achieve Good Ecological Status and were characterised as of moderate, poor and even bad ecological status. The exception was the case study on the Dhünn River (D) where the single water body of the study area was assessed as having Good Ecological Status and therefore no further assessment with regard to designation was undertaken. The latter case study showed a case where it was possible to achieve Good Ecological Status downstream of a reservoir in the vicinity of a dam. Also in the case of the Kennet River (UK), all water bodies appeared to be at Good Ecological Status. However, the authors of this case study considered that the assessment of biological status might be biased due to the fact that the biological samples in the study area were taken at sites of good habitat. Therefore, all water bodies proceeded to HMWB provisional identification.

In Table 14.15 (see Annex II), the main reasons and factors that led to lower than Good Ecological Status in the HMWB case studies are summarised. Selected factors, including physical alterations and hydromorphological changes were:

- In relation to hydropower:
  - Hydropeaking power operations;
  - Lake regulation for hydropower (high regulation amplitude);
  - Affected flow regime and flooding events mostly due to hydropower;
  - Sediment starvation and changes in vegetation downstream of catchwaters.

- In relation to other uses (flood protection, water supply, navigation, urbanisation, etc.):
  - Presence of impounded lakes affecting temperature and nutrient concentration;
  - Disruption of the river continuum due to weirs and dams as well as affected water temperature;
  - Construction of reservoirs and downstream river regulation;
  - Lack of estuary salinity gradient due to the construction of sluices;
  - Lake water level management for flood protection and/or navigation causing unnatural bank structure;
  - Biota degradation due to mill dams as well as minimum flow deficit;
  - Destruction/alteration of suitable biota habitats (bank and in-channel engineering, lack of floodplain and lateral connectivity, lack of meanders).

6.6 Conclusions and Discussion

The main points of discussion in the case studies are summarised.

Data and Methods for the Assessment of Current Ecological Status

Different combinations of biological, physicochemical and hydromorphological elements indicative of biological quality elements were used in the HMWB case studies for the purpose of ecological status classification.

Shortfalls in data were reported mainly for the *biological elements* (especially aquatic flora). The existing data and parameters from current monitoring systems were often judged as incomplete or not matching the requirements of the WFD and the HMWB designation. In some cases of considerable lack of appropriate biological data (as in the Finnish case study), indirect criteria based on physical parameters were suggested for the classification of the ecological status (and the subsequent provisional identification of HMWB).

The authors of the UK-EW case studies reported that there was huge potential for bias on ecological status, since impounded stretches in the UK-EW were not routinely sampled and existing monitoring schemes did not sufficiently cover potential Heavily Modified Water Bodies. Historically, the data had mostly been collected for water quality assessment purposes. This meant that the "best" habitats were generally sampled. The authors of these case studies noted that there was relatively little systematic sampling of the biology of different types of modification.

Many case studies reported that there was a lack of ecological assessment methods that integrate all quality elements of the WFD into a five-class system to evaluate ecological status. Examples were the lack of such a system for large rivers as well as lakes in the Netherlands and for lakes in Finland.

The authors of the case study on the Umealven River (S) suggested that hydromorphological parameters should be more widely used to assess ecological status. Nevertheless, several case studies also emphasised that knowledge was still rudimentary on the most important aspects of hydromorphological processes for the different organisms.

In some HMWB case studies, outside pressures (e.g. acidification, oceanic pressures on salmon) were difficult to *distinguish* from local *pressures* (e.g. regulation). It also appeared difficult to separate water quality impacts on biology from physical alteration impacts and hydromorphological change effects. This was particularly the case when different pressures affected the same parameters (e.g. the non-physical modification impact of water abstraction versus the impact of a physical barrier on flow). Therefore, in several case studies the need was expressed to adjust existing methods or develop new methods that can distinguish between the biological impacts resulting from different types of pressure.

A new biological quality element called "ecotopes" was proposed in the case study on the Haringvliet Estuary (NL). According to this case study, "ecotopes" are landscape ecological units and function as ecologically meaningful units holding specific sets of habitats for specific species of plants and animals. They have been proposed as a link between hydromorphology and ecology, which seems to have been poorly studied to date. The latter was identified as a common research need through the evaluation of the case studies and the discussions of the HMWB Working Group.

According to the final guidance document on HMWB and AWB, the assessment of the ecological status can be based directly on biology. Alternatively, indicative data (hydromorphological and physicochemical elements) can be used in situations where only these data are available (WFD CIS Guidance Document No. 4 2003).

Moreover, the final guidance document on HMWB and AWB notes that defining the extent of ecological damage in the manner required by the WFD will not be possible until common ecological monitoring is in place by 2006. Since the likelihood of failing to achieve GES should be assessed by 2004 (in time for the initial characterisation as in the WFD Art. 5), assessments may be estimates based on existing biological monitoring data and ecological classification systems.

Biological Quality Elements as Indicators for Physical Alterations

The suitability of several biological elements as indicators for physical alterations was regularly discussed in the case studies. In general, it was often cited that relationships between hydromorphological characteristics and ecological functioning of different groups of species are not well defined. Physical alterations have remained incidental to most monitoring programmes, and it was suggested by the HMWB Working Group that there is a *need to study the effects of physical alterations on biota in all water categories*. In the French case study on the Rhone River, it was argued that multiple factor interaction affects biological elements, making it difficult to establish a link between physical environment alterations and biological indicators. It was also argued that the use of biological indicators strictly linked to the water environment does not give a representative image of physical alterations and their ecological effects. The authors of this French case study noted that hydromorphological processes have visible repercussions on non-water sectors of valleys, which cannot be directly correlated to the status of water flora and most water fauna – except for fish more sensitive to physical alterations. Other indicators, e.g. concerning alluvial dynamics, would probably provide precious information, but a generalised collection of those would require a large-scale means and methodology.

The main points of discussion of the HMWB case studies with regard to biological quality elements as indicators for physical alterations can be summarised as follows:

- Benthic invertebrate fauna and fish were proposed as the most relevant groups for the assessment of *hydropower generation impacts*. In particular, according to the authors of the German case studies the number of fish species that have to migrate long distances compared to their number in a natural situation can serve as criteria for the assessment of hydropower impact;
- For *linear physical alterations* such as flood works, benthic invertebrate fauna and macrophytes/phytobenthos were proposed as the most appropriate indicators;
- Regarding specific regions such as the Nordic region, similarities in flora, fauna and water use make it possible to use the same indicators (sensitive species) for the assessment of impacts and the classification of ecological status.

Specific conclusions relevant to the *use of benthic invertebrate fauna* were the following:

- According to the case study on the Great Ouse (UK), the benthic invertebrate fauna groups of Coleoptera and Gastropoda have a high proportion of taxa associated with submerged and emergent aquatic plants. Therefore, these taxa may disappear or be reduced in case of degradation of these habitats due to physical alterations.
- Dragonflies, as assessed by the dragonfly-based assessment method used in Austria, are reliable indicators of ecological status with respect to lateral connectivity.
- In intertidal zones as in the Forth Estuary (UK), existing productivity estimates for benthic invertebrate fauna may be valuable in assessing losses due to land claim.
- The extent to which physical modifications impact upon benthic invertebrate fauna and other biota is not readily defined in UK rivers and streams at present. The current methods for the interpretation of macro-invertebrate data for rivers (BMWP/RIVPACS) do not permit separation of the influences of habitat quality and water quality. The authors of the UK-EW guideline report (Booker et al. 2002a) recommended the development of a new benthic invertebrate fauna scoring system, which would respond to physical alteration stress and can make use of the current and past benthic invertebrate fauna monitoring data. Preliminary assessments of the presence/absence of benthic invertebrate fauna taxa with well-defined habitat requirements indicate that a new system addressing the influences of various types of physical modifications could be developed.
- In the German case studies, the value of a new methodology called AQEM was pointed out. This method allows the establishment of relationships to specific pressures and impacts that cause changes to macroinvertebrate communities. The German case studies also suggested that an appropriate modified version of AQEM could be suitable for the assessment of GEP of HMWB with regard to macroinvertebrates. In this case, the reference conditions need to be redefined and single "metrics" modified.

Specific conclusions relevant to the *use of aquatic flora elements* were the following:

- In the case study on the Lozoya River (E), phytobenthos and especially macrophytes were proposed as good indicators of changes in flow downstream of reservoirs.
- According to the Swedish case studies, riparian plants respond to even small changes in flow regimes; therefore, they can be considered good indicators for physical alterations related to hydropower.
- According to the Finnish case study, macrophytes are suitable for the assessment of regulated lakes because they are sensitive to water level fluctuation. Such effects have been extensively studied in the Nordic countries, where riparian vegetation or littoral vegetation is sensitive to changes in the magnitude of spring flood. A specific indicator used in the Finnish case study was the growth area of the *Carex* species, which is directly related to water level fluctuations.

Phytoplankton was suggested by the authors of the Spanish case study as highly suitable for determining the trophic state of reservoirs. Apart from reflecting changes in catchment use and pollution, phytoplankton reflects the hydrologic regime of the reservoir. According to the authors of this case study, low phytoplankton density and concentration is indicative of the frequency of water renewal.

According to the final guidance document on HMWB and AWB, the biological status of surface water is to be assessed using the appropriate elements in the different water categories (WFD Annex V 1.1) (WFD CIS Guidance Document No. 4 2003). The final guidance document suggested that the preliminary assessment of the ecological status, to be completed by 2004, should be based on the most sensitive quality elements with respect to the existing physical alterations. The document noted, however, that this procedure concentrates on the *effects* of physical alterations on some sensitive elements of the aquatic ecosystem.

The final guidance document on HMWB and AWB also mentioned that to detect the reason for the possible failure of the environmental objective (i.e. the good status or potential), indicative parameters differ according to the causes (WFD CIS Guidance Document No. 4 2003). The guidance document on HMWB and AWB is particularly concerned with indicative data to detect hydromorphological changes. Effects resulting from other impacts (e.g. toxic effects on macroinvertebrates, eutrophication concerning macrophytes) should be differentiated as far as possible (WFD CIS Guidance Document No. 4 2003). Some suggestions of the final guidance document on the suitability of biological elements as indicators for physical alterations have been based on the above conclusions.

The Definition of Ecological Status in the WFD

According to the authors of several of the HMWB case studies, it was difficult to define the limits between the WFD classes of ecological status due to the vague definitions of the WFD.

In summary, as concluded by the UK-EW, there was no scientifically robust analysis that could be used to translate the WFD definition of ecological status into quantitative figures. However, the normative definitions of ecological status classifications found in the Water Framework Directive provide a clear qualitative classification of high, good, moderate, poor, and bad. It was suggested that this type of approach allows for freedom of interpretation and will produce discrepancies in the classification of ecological status due to the variability found between different experts' opinions.

The authors of the Spanish case study commented that the definition of ecological status in the WFD is weak since it gives ground for an assessment of subjective nature. It was questionable as to how results from different indicators would be combined and whether higher importance should be given to a specific group of organisms for the assessment of ecological status.

In the case of the Suldalslagen River (NO), physical alterations did not lead to a loss of species but to an alteration of the abundance and balance between species. It was chosen at that stage to define "poor status" as "loss of species" and to use "moderate status" for "large changes in abundance". However, the authors of this case study recognised that further development of the ecological status classification criteria might change the latter definitions.

Reference Conditions to Assess Ecological Status

From the evaluation of the HMWB case studies it was made obvious that the selection of the reference condition is a crucial step in the HMWB provisional identification process. The definition of the reference condition itself should be clarified, including whether it should be a historical or a modelled situation. It was also unclear whether reference conditions refer only to pristine waters or also to waters under human influence.

In several cases, it was difficult or impossible to define a reference condition, making it therefore difficult to gain a true picture of GES. For rivers that had been modified for a very long time, determination of any reference state was problematic.

Instability of Ecological Status

The issue of instability of the ecological status was discussed in two case studies. In the case of Hagmolenbeek (NL), the values for ecological status differed from year to year, as they depended on the amount of precipitation that had fallen in a certain year. In wet years, the ecological condition was better than in dry years. To overcome this problem, the average result over a period of ten years was used to classify ecological status. In the case of the Suldalslagen River (NO), the water body was not in a stable condition but in a transitional state due to the changes in the hydrological regime caused by river regulation. It was pointed out that the classification for the purposes of the case study referred to the present state of the water body and not to a foreseen stationary situation in the future.

Chapter 7

Provisional Identification of Water Bodies as Heavily Modified

7.1 Introduction

As shown in the identification and designation process for Heavily Modified and Artificial Water Bodies (HMWB and AWB) (Step 6 of Fig. 2.1, p. 12), those water bodies should be selected as provisionally HMWB where the changes in the hydromorphology result in the water body being substantially changed in character. The remaining water bodies, which are likely to fail GES but are not substantially changed in character due to hydromorphological changes, will not be identified as HMWB. The provisional identification of water bodies as HMWB should be completed by 22 December 2004 and is part of the characterisation of River Basin Districts as defined in Annex II of the WFD.

In this chapter, the main results, approaches and factors on which the decision as to whether a water body was provisionally identified as HMWB are summarised.

7.2 Results

A clear outcome on the identification of water bodies as provisionally HMWB was presented in twenty-seven case studies (see Table 14.15 in Annex II for results). In the remaining seven case studies, no conclusion on HMWB provisional identification was explicitly mentioned although all or some of the water bodies were further assessed with regard to the designation tests, implying that they had also been provisionally identified as HMWB.

In eleven case studies, all water bodies within the study area were identified as provisionally Heavily Modified Water Bodies. In the remaining cases, some water bodies were provisionally identified as HMWB while the rest were considered as natural. Factors that contributed to the provisional identification as Heavily Modified Water Bodies coincided in most cases with the reasons for failure to achieve Good Ecological Status (see Table 14.15 in Annex II).

7.3 Criteria and Methods

The approach followed in most case studies to reach a decision on provisional identification as HMWB was more or less compatible with the approach given in Fig. 2.1 (p. 12). This involved going through the individual steps leading to provisional identification via the assessment methods that have been described in Sects. 5.3, 5.4, and 6.5 of this book.

More explicitly, in most of the case studies, provisional identification of HMWB was based on the existence of physical alterations and hydromorphological impacts as well as the failure to achieve Good Ecological Status. In five case studies (Haringvliet Estuary (NL), Umealven River (S), Forth Estuary, and rivers Dee and Lagan (UK)), the decision on provisional identification as Heavily Modified Water Bodies was based only on the failure to achieve Good Ecological Status.

In some cases, more elaborate procedures slightly deviating from that presented in Fig. 2.1 (p. 12) were used for reaching a decision on provisional identification as HMWB. The main ones were the following:

- In the case study on the Sarre River (F), a decision-making procedure consisting of five successive questions was used to structure the identification and designation of Heavily Modified Water Bodies and their assessment. Stage 1 and Stage 2 of Fig. 7.1 led to the identification of HMWB "candidate" water bodies (provisional identification) and consisted of the following two main questions:
  - Question 1 – *Has the section been significantly modified in an obviously and not easily reversible way?* This was termed the "hydro-morphological filter" that was used to select pre-candidate sections on the basis of physical criteria. The degree to which the control variables of the physical environment (stream discharge and sediment discharge, riparian vegetation and valley gradient) have been modified was estimated using the response variables available (evolution of the fluvial style, mean flow) (see Sect. 5.3 of this book for a description of this stage). If one of these control variables has been significantly modified, the section was selected for verification of its biological status.
  - Question 2 – *Does the ecological status of the pre-candidate section fall below the "good" category?* This was termed the "biological filter", which was used to select candidate sections (provisionally HMWB) on the basis of ecological criteria. The four groups of fish fauna, macrophytes, phytoplankton and invertebrates, with the inclusion if necessary of "additional physical indicators" were graded according to the type of water course. A weighting system based on the number of taxon species and existing assessment indices was proposed to evaluate the ecological status of the sections. Degradation in just one of these biological groups is enough to qualify the section as a provisionally HMWB.
- According to the case studies of the UK-EW, consultation on the application of the designation tests requires a transparent and full operation of the designation tests for each water body. In cases where there are many provisional HMWB, this can be very time-consuming. The authors of these case studies argued that some of the designations can be contentious and require good, well-argued, and complex designation cases. Other cases may be less likely to be disputed, as the water bodies are obviously HMWB or obviously not HMWB. Therefore, it was suggested that a screening approach is needed, which involves removing those water bodies that are clearly HMWB and those that are clearly not HMWB in order to concentrate effort on more borderline cases (for details on the entire UK-EW procedure see also Sect. 8.4 of this book). It was also proposed that provisional identification should be done as soon as possible to give time to operate the designation tests for a large number of water bodies and allow for consultation. The proposed screening followed the pre-screening stage (for details on pre-screening see Sect. 5.3 of this book), providing a rapid

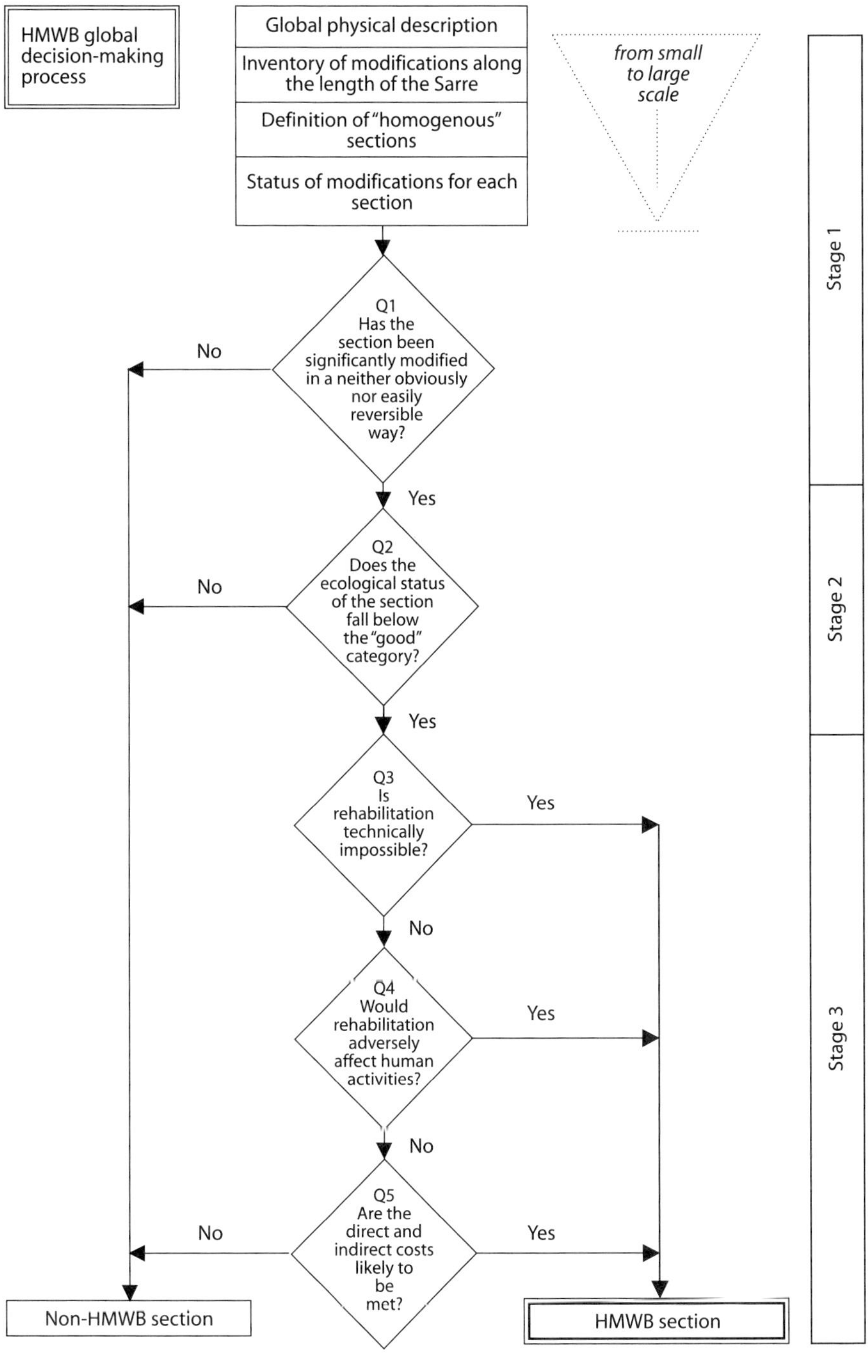

Fig. 7.1. Decision-making diagram for the identification and designation of HMWB, France. *Source:* Case study on the Sarre River, F

method to decide whether GES could be reached. If a potentially or borderline Heavily Modified Water Body as defined through the pre-screening method also failed to achieve Good Ecological Status, it was provisionally heavily modified and proceeded to the designation tests.

- In the case study on the Dender River (B), water bodies were identified as provisionally heavily modified by plotting the intensity of physical modifications against the values on ecological status. The boundary between HMWB/not HMWB was set by expert judgment by determining threshold values. The following rule was set: If the degree of modification was higher than 40% and the value for ecological status was less than 60 (numbers refer to the sum of the different indicator values expressed as percentages), the water body was provisionally heavily modified.
- In three German case studies (rivers Elbe, Lahn and Seefelder Aach), a three-step process based on the development of *positive/negative lists of physical alterations* and hydromorphological impacts was used for provisional identification as follows (example given in Fig. 7.2):
  - Step 1: characterisation of pressures as significant or not significant according to the LAWA *(Länderarbeitsgemeinschaft Wasser)* criteria (for details on the LAWA criteria refer to Sect. 5.3 of this book);
  - Step 2: evaluation of ecological status (failure to achieve good status or not);
  - Step 3: separation of the impacts of a significant pressure upon hydromorphology and biology into negative and positive lists according to the findings of Steps 1 and 2.

  The aim of this approach was to simplify the HMWB provisional identification (and later designation) process by defining "non-significant" physical alterations (positive list) and "significant" physical alterations (negative list). A positive list means that for these physical alterations, measures can be identified to achieve Good Ecological Status. A negative list includes physical alterations that influence the biology so strongly that the water body must be identified (and later designated) as HMWB. Therefore, the development of a negative list leads to the provisional identification of the water body as HMWB. A positive list and a negative list for the provisional identification of water bodies affected by navigation are presented in Box 7.1.
- In the case study on Lake Kemijärvi (SF), the approach followed for provisional identification was in practice based on the *indirect criteria of the REGCEL water level analysis* (e.g. changes in water levels, flows that are critical for biota) due to the limited biological data (for more details on the method REGCEL see Sect. 5.4 of this book). In this case, biological data were utilised wherever possible. As part of a large project aiming to develop criteria for the identification of heavily modified lakes, the REGCEL water level analysis was applied to a large number of lakes in Finland (with regulated and non-regulated water levels). Thereby, the whole set of Finnish lakes was divided into three groups based on water level fluctuation: (*i*) natural lakes, (*ii*) probably heavily modified lakes and (*iii*) heavily modified lakes.

The final guidance document on HMWB and AWB suggests that if a water body is to be provisionally identified as heavily modified, the following criteria apply (WFD CIS Guidance Document No. 4 2003):

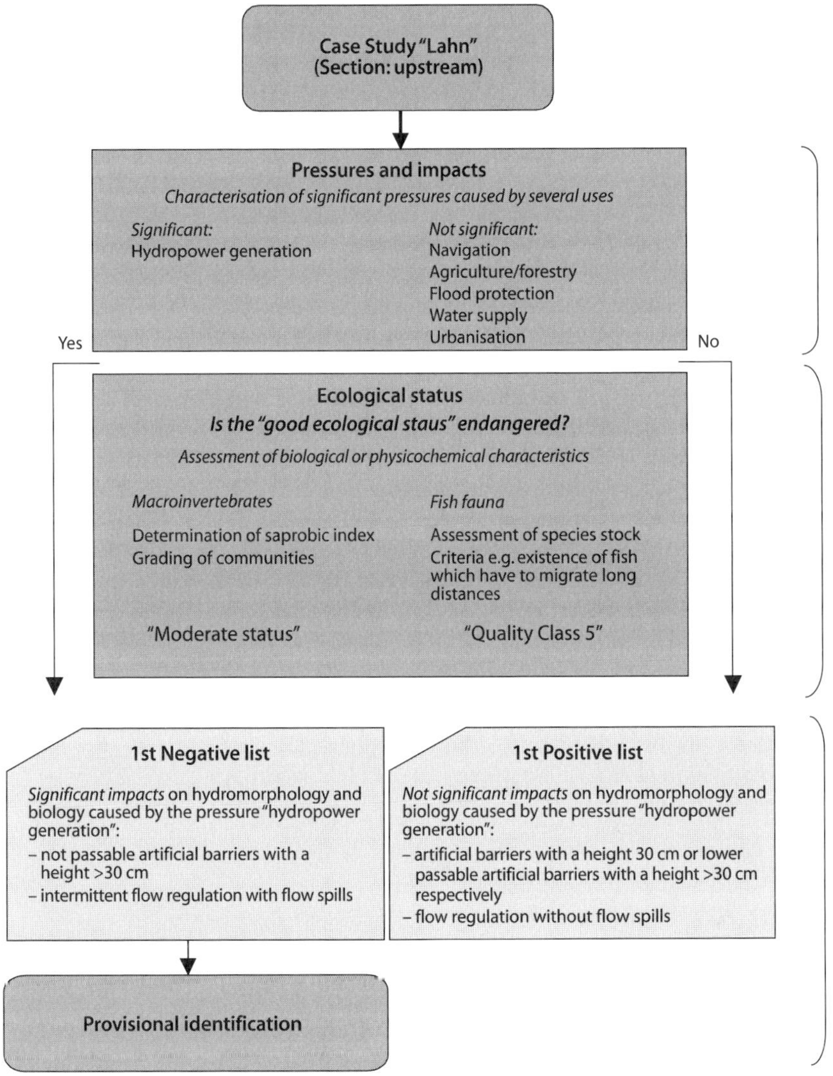

Fig. 7.2. Process of the "Provisional Identification of Heavily Modified Water Bodies", Germany. *Source:* Case study on the Lahn River, D

- The failure to achieve good status results from *physical alterations* to the hydromorphological characteristics of a water body. It must not be due to other impacts, such as physicochemical impacts (pollution).
- The water body must be *substantially changed in character*. This is the case when there is a major change in the appearance of the water body. It is clearly a partly subjective decision as to whether a water body is (*a*) only *significantly* changed in

Box 7.1. Positive and Negative Lists of Physical Alterations for the HMWB Provisional Identification of Waters Affected by "Navigation"

Positive list of physical alterations:

- 10% impounded river length at mean low water flow;
- Proportion of river length with discharge acceleration with
  - Ratio profile depth to profile width <1:4; or
  - Bank (single or both sides) <10% total length with bank impairments; or
  - Longitudinal profile <70% stretched or straightened;
- Artificial barriers with a height = 30 cm, passable artificial barriers with a height >30 cm respectively;
- Cross-linking of the river with oxbow lakes existent;
- Dredging.

Negative list of physical alterations:

- Bank reinforcement, dikes (>10% impounded river length at mean low water flow);
- Migration of fauna not possible (not passable artificial barriers with a height >30 cm);
- Channelisation (longitudinal profile = 70% stretched or straightened);
- Cross-linking of the river with oxbow lakes non-existent;
- Harbour structures, port facilities;
- Reduced stream flow velocity;
- Reduced tidal amplitude in estuaries;
- Turning of an estuary into a freshwater reservoir.

Source: Material agreed during the meetings of the HMWB case study subgroup on navigation.

character (e.g. water abstraction without morphological alterations) or (*b*) *substantially* changed in character when provisional identification as HMWB may be appropriate (e.g. long-term hydromorphological changes caused by a weir). Both may be likely not to achieve GES. However, the following considerations should be borne in mind:
  - When visiting a water body that is substantially changed in character, it should be very obvious that the water body is substantially changed from its natural condition;
  - The change in character must be extensive/widespread or profound. Typically this should involve substantial change to both the hydrology and morphology of the water body;
  - The change in character must be permanent and not temporary or intermittent.
- Many alterations to the hydrological characteristics of water bodies, such as abstractions and discharges, are not associated with morphological changes, and may therefore often be easily reversible, temporary or short-term. Consequently, such alterations would not constitute substantial changes in the character of water bodies and hence the application of HMWB designation would not be considered.
- The modification must be consistent with the scale of change that results from the activities listed in Article 4(3)(a) of the WFD: a canalised river, a harbour, a river constrained for flood protection or a dammed river or lake.
- The substantial change in character must be the result of the specified uses. It must have been created by uses listed in Article 4(3) of the WFD or uses which represent equally important sustainable human development activities (either singly or in combination).

7.4
Scope and Extent of Provisional Identification

This section summarises the case study approaches on the scope and extent of provisional identification as HMWB. The support material (guidance papers) agreed by the HMWB WG for conducting the case studies identified two options for the scope and extent of provisional identification:

a *Provisional identification as HMWB on the basis of physical alteration (morphology)*
Only the river section directly physically altered, e.g. by the construction of a dam is provisionally identified as HMWB. The water bodies upstream and downstream are not physically altered themselves, but they are influenced by the HMWB. Consequently, "less stringent environmental objectives" may be set for the upstream and downstream water bodies.

b *Provisional identification as HMWB on the basis of changes to the hydromorphological characteristics (hydromorphology)*
The whole area affected by the physical alteration is provisionally identified as HMWB. The HMWB includes the full part of the river where the ecology is affected and not only the stretch directly altered physically.

A clear distinction between these two options could be made in cases where the water body was affected by physical alterations that interrupted the river continuum (e.g. dams with large impoundments). On the contrary, the distinction was less clear in cases of physical alterations that did not interrupt the river continuum such as river straightening or bank reinforcement.

The two options were assessed in seven case studies (Lozoya (E), Kemijärvi (SF), Hagmolenbeek (NL), Emån (S), Umealven (S), Tummel (UK), and Nestos (GR)). All except for the case study on the Hagmolenbeek River dealt with the specified use of "hydropower generation". In the case studies on the rivers Hagmolenbeek (NL) and Emån (S), preference was expressed for option *(a)*. Use of option *(a)* led to a smaller number of provisionally identified HMWB. On the other hand, provisional identification on the basis of hydromorphology (option *(b)*) could lead to water bodies being unnecessarily designated as heavily modified, thereby reducing the environmental objectives to be achieved. Moreover, it proved easier to identify the extent of morphological changes as opposed to hydromorphological changes.

The remaining five case studies ((Lozoya (E), Kemijärvi (SF), Umealven (S), Tummel (UK), and Nestos (GR)) supported option *(b)*, arguing that a physical alteration such as a dam affects the ecological status of upstream stretches as well, especially regarding migration continuity and water body type. Furthermore, any restoration measure would inevitably change the hydromorphology of the adjoining water bodies.

The authors of the German case studies also suggested that provisional identification may include a division of water bodies according to different morphological alterations (different HMWB types). In view of expected adverse effects of the restoration measures on the uses, the case study authors recommended to further subdivide provisionally identified HMWB into sub-bodies. This means that the boundaries of the initially identified water bodies can be adapted after the findings of the provisional identification process, implying that the initial water body identification should be an iterative process.

With relevance to this, the final guidance document on HMWB and AWB recognised that within the provisional HMWB identification, the scale, scope and extent of water body identification should be considered. It may be necessary to adapt the boundaries of the initially identified water bodies according to the substantial changes in hydromorphology. More specifically, where the hydromorphogical changes do not coincide with the boundaries of a surface water body, it may be appropriate to subdivide the water body in order to separate heavily modified stretches from the unaffected areas of the water body (WFD CIS Guidance Document No. 4 2003).

7.5 Conclusions and Discussion

Different approaches for reaching a decision on the provisional identification of HMWB were encountered in the case studies. Provisional identification was based mainly on:

- The existence/evaluation of physical alterations and hydromorphological impacts as well as the result of the classification of current ecological status;
- Screening approaches as in the case of UK-EW;
  - A screening approach may assist in directing efforts to complex borderline cases of HMWB designation. It should be possible to screen out early in the process obviously HMWB or obviously not HMWB.
- Threshold values on the degree of modification and ecological status set by experts;
  - In the Belgian case study, the threshold values determined whether a water body was provisionally HMWB or not.
- Positive and negative lists of physical alterations (German case studies);
- Physical parameters (indirect criteria) due to limited biological data (as in the Finnish case study).

With regard to the latter, the German case studies argued against the use of indirect criteria for provisional identification. The authors of these case studies concluded that the final decision on provisional identification (and designation) as HMWB should not be based on morphology and pressure identification alone. The biological status should be the decisive factor. This argument was supported by the findings of the case study on the Dhünn River (D), where a water body was identified as natural even though it was downstream of a reservoir in the vicinity of a dam. Despite the morphological alterations, Good Ecological Status was possible to be achieved. Therefore, the authors of the German case studies recommended a specific analysis of the local conditions for each case.

The HMWB case studies identified the need for further discussion on the following issues, which remained unresolved during the course of the studies:

- The combined impacts of water quality and hydromorphological changes on ecological status may confuse the provisional identification of HMWB as in the cases of the rivers Tame and Sankey Brook (UK). In these case studies, it was concluded that the water quality issue may be a problem for assessing many HMWB reaches. According to the case study authors, poor water quality results in difficulty with assessing the biological impacts of physical alterations and thus uncertainty as to whether HMWB identification and designation is appropriate.

- The authors of the Finnish case study identified the need to clarify the importance of the HMWB provisional identification. It is thereby important to clarify if and how the number of provisionally HMWB (either under- or overestimated) would influence the final result of designation.
- It also appeared necessary to define how water bodies with disturbed river continuum and loss of migratory fish species due to barriers in downstream stretches should be handled.

Regarding this issue, which is linked to the scope and extent of HMWB provisional identification (see Sect. 7.4 above), the German case studies suggested that water bodies, which have a "good status" according to the biological quality elements except for anadromous fish, should not be designated as HMWB but as having "less stringent environmental objectives" (as in the case of the upper Lahn River and Dhünn River, D). The authors of the case study on the Tummel River (UK) came to the same conclusion, arguing that "less stringent environmental objectives" should be required for those water bodies upstream of hydropower schemes where migration of species has been compromised by structures built as part of these schemes. In the Dee River case study (UK), it was also argued that if the upstream section is not physically modified, it should not be considered for HMWB provisional identification but for "less stringent environmental objectives".

The final guidance document on HMWB and AWB (WFD CIS Guidance Document No. 4 2003) actually clarified this point. It is important that only water bodies that are substantially changed in character (due to physical alterations) themselves may be provisionally identified as HMWB. If a physical alteration (e.g. dam) impacts the biological quality elements in the upstream part of a river system (for example fish migration is hindered), this upstream part may not be considered for provisional HMWB identification. If the GES cannot be achieved in this water body upstream of a physical alteration, the environmental objective may be less stringent (WFD CIS Guidance Document No. 4 2003).

Chapter 8

Designation of Heavily Modified Water Bodies

8.1
Introduction

According to the identification and designation process for Heavily Modified and Artificial Water Bodies (HMWB and AWB) (Fig. 2.1, p. 12), the provisional identification of HMWB is followed by the designation process for HMWB (Steps 7 and 8 of Fig. 2.1). Once a water body has been identified as HMWB, it may be designated as such if according to the WFD Art. 4(3)(a) and (b):

a “The changes to the hydromorphological characteristics of that body which would be necessary for achieving Good Ecological Status would have significant adverse effects on:
   i the wider environment;
   ii navigation, including port facilities, or recreation;
   iii activities for the purpose of which the water is stored, such as drinking water supply, power generation or irrigation;
   iv water regulation, flood protection, land drainage; or
   v other equally important sustainable human development activities”.[1]

   and

b “The beneficial objectives served by the artificial or modified characteristics of the water body cannot, for reasons of technical feasibility or disproportionate costs, reasonably be achieved by other means, which are a significantly better environmental option.”

Figure 8.1 presents in detail the relevant steps required to address Art. 4(3)(a) and (b). The designation process hence consists of two main steps, the “designation test 4(3)(a)” and the “designation test 4(3)(b)” (see above). A water body may be designated as heavily modified, if it has passed through the designation procedure involving both designation tests.

In this chapter, the results and different approaches used in the HMWB case studies with regard to the application of these designation tests are presented.

[1] Terms included under points (ii–v) of the WFD Art. 4.3(a) are referred to as “specified uses”.

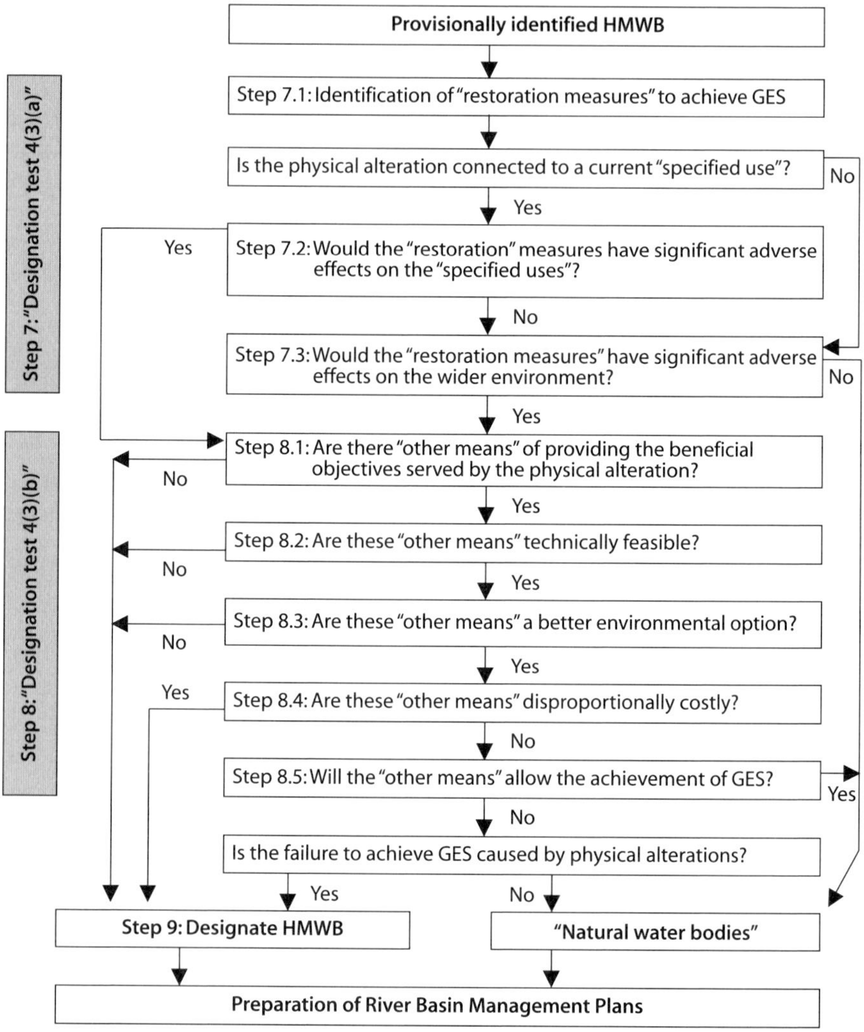

Fig. 8.1. Steps leading to the designation of HMWB (Art. 4(3)). *Source:* WFD CIS Guidance Document No. 4, 2003

8.2 Designation Test 4(3)(a): Assessment of Necessary Restoration Measures to Achieve GES

Designation test 4(3)(a) is the first step of the designation process after the water bodies have been provisionally identified as heavily modified (see Fig. 8.1). This test identifies whether the necessary hydromorphological changes to achieve GES (also referred to as "restoration measures" within the HMWB Working Group) have a significant adverse effect on the specified uses such as hydropower or navigation or upon the wider environment, e.g. nature reserves.

8.2.1 Identification of the Necessary Restoration Measures

In this section, reference is only made to the necessary *changes to hydromorphology* (restoration measures) to achieve GES. In the HMWB case studies, these measures ranged from measures aimed at reducing the environmental impact of the physical alteration (e.g. via increased compensation flows or fish passages) to measures resulting in the complete removal of the physical alteration (Table 14.22 see Annex II).

In cases of obvious physical alterations such as dams or weirs, the identification of necessary hydromorphological changes was relatively straightforward. The identification of restoration measures was more complex in cases of less obvious physical alterations and mixed uses (Great Ouse, Sankey Brook (UK)) as well as long-term historical modifications (Hagmolenbeek (NL)).

The identification of the necessary restoration measures to achieve GES was mainly based on expert judgment. In only a few cases, the identification of measures was based on more concrete methodologies as in the case of the Lozoya River (E); in this case, measures were defined according to the ecological flow requirements for maintaining optimal fish fauna biomass (applying the method PHABSIM, Physical Habitat Simulation). In the case study on the Sarre River (F), proposals for restoration measures were made by assessing the link of biological deterioration to physical modifications. From the analysis of the HMWB case studies, two broad categories of restoration measures were identified:

1. Restoration measures directly related to the existing physical alteration. This approach usually resulted in a change or removal of the physical alteration (e.g. removal of embankments, removal of weirs, lowering of bottom barrier of a dam) (examples can be in found in the case studies on Dender (B), Haringvliet (NL), Lozoya (E), Beiarn (NO), and Umealven (S)).
2. Restoration measures that were not necessarily related to single physical alterations but rather enhanced the general ecological conditions (e.g. creation of habitats, re-establishment of dynamic floodplains, lower land use intensity, and establishment of alluvial forests) (examples can be in found in the case studies Danube (A), Seefelder Aach (D), and Ruhr (D)).

The authors of the case study on the Ruhr River (D) suggested that with regard to the identification of restoration measures, a cause-and-effect analysis is essential in order to derive the necessary restoration measures to achieve GES. This analysis has to focus on actual pressures and the resulting ecological deficits. It is assumed that the deficits can be remedied by different measures. To make the decision process transparent, the case study authors recommended developing different scenarios. The scenarios and their measures should be oriented towards achieving GES (for relevant examples from the German case studies, see Table 14.22 in Annex II). Under no circumstances should the scenarios be developed considering most of the uses as indispensable. Restrictions should be made only for "clear-cut situations". The overall procedure proposed can be summarised in three steps:

- The impacts and resulting effects (physical and biological) are listed for each water body;
- Subordinate targets are developed (e.g. reduction of eutrophication, reestablishment of river continuum, restoration of natural stream bed); and
- Different scenarios (solution alternatives) of measures are developed.

During discussions, the HMWB Working Group (WG) concluded that initial check lists of restoration measures to improve hydromorphology would be helpful for different specified uses. Essentially this could be "best environmental practice" guidance for particular types of uses. The lists of restoration measures shown in Tables 8.1 and 8.2 were presented by the HMWB WG joint chair and agreed upon at a workshop on HMWB on 30–31 May 2002 in Berlin.

Effectiveness of Restoration Measures

In many case studies, some indication was made of the expected positive effects of the proposed restoration measures (refer to relevant column of Table 14.22 in Annex II). Actually, the search for measures (and their scale of effects) that aimed to reduce the environmental impact of the physical alteration proved to be more difficult than for measures which resulted in the removal of the physical alteration. This was because the effects of each restoration measure as well as of their cumulative effect needed to be predicted in order to assess measures (or sets of measures) that lead to GES. For these predictions, methods of ecological modelling using empirical values were used (as in the case study on Lake Kemijärvi (SF), the Hagmolenbeek-Hegebeek River (NL) and the Haringvliet Estuary (NL)). Similarly, in the case study on the Ruhr River (D), the targeted volumes of the impounded lakes in question were calculated based on the reproduction behaviour of algae and flushing effects.

The Dutch case study on the Hagmolenbeek River commented that in the context of the HMWB designation process, an exact cause-effect-relationship between a measure and the resulting ecological condition was assumed. According to the authors of this case study, such a relationship is for most ecosystems very difficult to derive and is accompanied by a high degree of uncertainty. In this specific case study, it was attempted to predict the restoration measures needed to achieve the GES by comparing the Hagmolenbeek with a reference stream with regard to hydromorphology, water quality and ecology. The case study authors noted, however, that this approach was debatable because two streams are never completely the same and can be comparable only to a certain extent.

The case of the Bregenzerach (A) was unique since the restoration measures proposed were already in place; this made it possible to give information on the actual effectiveness of the measures with regard to the achievement of GES.

In cases where poor ecological status resulted from both physical alterations and low water quality, the effect of restoration measures on water quality and the ecological status were nearly impossible to differentiate using current methodologies and knowledge (e.g. in the cases of Sankey (UK) and Tame (UK)). Furthermore, in some

Table 8.1. Restoration measures for hydropower. *Source:* Material presented at the HMWB workshop on 30–31 May 2002 in Berlin

Physical alterations	Restoration measures
Water flow and water-level fluctuations	Reintroduce a spring flood Decrease flow during summer to winter Avoid flash floods downstream dams Introduce minimum flows
Habitat improvements	Terrace tributary outlets (reservoirs) Reduce erosion in riparian zone Return boulders to channel (former rapids) Seed and plant species
Disruption in river continuum	Remove dam Build in fish passages, improve existing fish passages Co-ordinated spillway releases (multiple dams)

Table 8.2. Restoration measures for navigation. *Source:* Material presented at the HMWB workshop on 30–31 May 2002 in Berlin

Physical alterations	Restoration measures
Dams and weirs	Build in fish passages, improve existing fish passages
Channel maintenance/dredging	Reduce intensity of dredging
Channelisation/straightening	Connecting existing meanders to the main channel Initiating meanders
Bank reinforcement	More natural embankments
Detachment of oxbow lakes and wetlands	Cross-linking of the river by deepening the oxbow lakes Establishment of natural floodplains

case studies (Suldalslagen (NO), Mulde (D), Dee (UK) and the Baltic coast (S)), it was argued that the effect of a restoration measure may be restricted by regional water quality problems such as acidification. In the case of the Suldalslagen River (NO) and the Dee River (UK), the recovery of fish population (salmon) might be hampered by acidification, even if migration continuity would be ensured.

According to the agreed approach of the final guidance document on HMWB and AWB (Step 7.1 of Fig. 2.1, p. 12), the restoration measures for achieving GES may range from measures aimed at reducing the environmental impact of the physical alteration (e.g. increased compensation flows or fish passages) to measures resulting in the complete removal of the physical alteration (see Sect. 5.4.1 of the WFD CIS Guidance Document No. 4 2003). Measures can be directly related to the physical alteration (e.g. changing the physical alteration), or they can enhance the general ecological conditions (e.g. creation of habitats). The contribution that an individual measure could make towards achieving GES needs to be predicted. It should also be assessed whether an overall package of proposed restoration measures could lead to GES. The measures should be well defined (e.g. exact percentage of compensation flow) and should include an assessment of whether GES status will be delivered (full or partial delivery). Combinations of "partial" measures may allow GES to be achieved. The identification of suitable measures can be difficult, because information on the cause-effect relationship of measures is often not sufficient. Finally, according to the final guidance document, the costs of restoration measures should not be considered in designation test 4(3)(a) (WFD CIS Guidance Document No. 4 2003).

8.2.2 Assessment of Significant Adverse Effects on the Specified Use or the Wider Environment

After the identification of restoration measures to achieve GES, it has to be assessed whether those measures do or do not have a significant adverse effect on the specified uses or upon the wider environment. If these measures *do not* have a significant adverse effect on the specified uses or the wider environment, the water body *should not* proceed to designation test 4(3)(b) and is considered natural (see Fig. 8.1).

Significant Adverse Effects on the Specified Uses

The majority of restoration measures in the HMWB case studies were considered to have significant adverse effects on the specified uses (see Table 14.22 in Annex II). Significant adverse effects included:

a *Complete loss of use:*
   - Loss of water supply (Lozoya (E), Haringvliet Estuary (NL));
   - Loss of hydropower production (e.g. Emån (S), Umealven (S));
   - Abandonment of navigation, e.g. due to lowering of water levels especially in the summer (Dender (B), Great Ouse (UK), Forth Estuary (UK), Baltic (S));
   - Loss of recreational facilities (Loosdrecht, Veluwerandmeren (NL));
   - Loss of flood protection (Great Ouse (UK)).

b *Reduction of use or increase of risk to lose the beneficial objectives of use:*
   - Partial take up of space used for urbanisation and agriculture (Dender (B));
   - Limitation of passenger navigation (Ruhr (D));
   - Effects on inhabited areas through flooding (Great Ouse/Wissey (UK), Loosdrecht (NL));
   - Increased flood risk (Dender (B), Authie (F), Kemijärvi (SF), Veluwerandmeren (NL), Forth Estuary (UK)).

c *Production (or socio-economic) losses:*
   - Reduced agricultural production (Haringvliet Estuary (NL), Veluwerandmeren (NL), Hagmolenbeek(NL), Forth Estuary (UK));
   - Reduced hydropower production (Danube (A), Kemijärvi (SF), Suldalslagen (NO), Beiarn (NO), Emån (S), Forth Estuary (UK), Tummel (UK), Dee (UK));
   - Loss of employment (Authie (F), Forth Estuary (UK)).

In two cases (Hagmolenbeek (NL), Sankey Brook (UK)), it was unclear whether the adverse effects of the proposed measures were significant. In both cases, the adverse effect was increase of flood risk in the adjacent area.

In a few cases, it was concluded that some or all measures would have no significant adverse effects. In the case of the Dender River (B), for instance, there would be no significant adverse effect of erosion control measures on agriculture.

In the German case studies on the rivers Elbe, Seefelder Aach and Lahn, the consideration of the adverse effects on different levels played a role in the assessment of their significance (see Table 14.22 in Annex II). In the case of the Elbe, the effect of decreased navigable depth (by 10–30 cm) on navigation was assessed on a regional level and was considered not to be significantly adverse. To reach this conclusion, the share of goods transported by cargo ships in the area of the Elbe (only ca 1%) was compared with the average share of goods transported by cargo ships in Germany (12%). Given the fact that this share in the Elbe is clearly below the German average, inland navigation on the Elbe was considered of secondary importance.

In the case study on the Seefelder Aach River (D), it was argued that an assessment of adverse effects on agriculture on a local level focuses on the economics of a single farm and requires detailed information such as the percentage of affected land in relation to the total farm size, farm structures, etc. Considering the difficulty in collecting such information, focus on a local level in the case of agriculture was not recommended by the case study authors and significance of effects was assessed on a regional level.

However, the German case studies were an exception in considering levels other than local. In the majority of the HMWB case studies, the scale for considering significant adverse effects was usually local, e.g. only the production loss at one hydropower station was considered rather than the overall national production loss.

In general, the interpretation of the term "significant" within the case studies has been based largely on expert judgment. With regard to the methodology used for the assessment of significant adverse effects, categories (*a*: complete loss of use) and (*b*: reduction of use) of adverse effects mentioned above were regarded as "clear-cut situations" and were usually assessed qualitatively.

The significance of adverse effects for category (*c*: production losses) was assessed with rough/single estimations of the loss of production (hydropower, yield of crops) and in a few cases it was calculated as economic loss (Haringvliet Estuary (NL), Hagmolenbeek, (NL)). The range of the significance of production losses differed widely, and a short overview is given in Table 8.3.

Production losses <5% were usually considered as not significant if no other additional adverse effects occurred. Production losses >30% were clearly identified as significant, without further quantitative assessment.

According to the discussions of the HMWB WG, indicators of significant adverse effects on hydropower may include the absolute cost of the effect, % of the hydropower scheme power output, % of the hydropower scheme revenue as well as the number of jobs at risk[2]. It was not considered possible to derive a standard definition for what a significant adverse effect is due to:

[2] From the material agreed during the meetings of the case study subgroup on hydropower.

Table 8.3. Values the for significance of adverse effects on the use (production losses). *Source:* Compiled by the authors from the evaluation of the HMWB case studies

Case study	Use	Significant	Not significant
Danube (A)	Hydropower	>50% local production loss	–
Kemijärvi (SF)	Hydropower	>30% local production loss	–
Beiarn (NO)	Hydropower	>15% local production loss	–
Emån (S)	Hydropower	>30% local production loss	–
Haringvliet (NL)	Agriculture	> yield reduction of 6 billion Euro	–
Hagmolenbeek (NL)	Agriculture	Yield reduction by 2 times ⇒ 1.8 million Euro yr^{-1}	≤5% loss of agricultural land
Forth Est. (UK))	Agriculture	>34% reduction of agricultural land	–
Seefelder Aach (D)	Agriculture	–	≤10% loss of agricultural land

- Differences in the value of electricity within Member States; and
- Differences in the contribution of hydropower to the national output within Member States.

In the case study on the Ruhr River (D), it was proposed that in order to conclude on the significance of adverse effects on the uses, it was necessary at first to develop criteria of significance. The definition of criteria of significance was partly dependent on the definition of the beneficial objectives of the regarded uses. For instance, in the case of agriculture, there could be two different beneficial objectives: production of food as well as maintenance of agricultural employment. Depending on the kind of beneficial objective considered, the criteria and thresholds of significance could differ. For this reason, the beneficial objectives should be defined sectorally as well as individually. According to the authors of this case study, the same use could have different beneficial objectives by sectoral or individual regard. As a second step, the adverse effects on the uses were qualitatively described, and finally the significantly affected uses were identified based on the levels of significance.

Some case studies also considered the costs of the restoration measures as part of designation test 4(3)(a) and even excluded expensive restoration measures from further assessments (as in the case study on the Emån River (S) regarding the costs for fish passages). In the case study on the Sarre River (F), the total budget for achieving GES was estimated to reach 4 625 000 Euros. A more extensive discussion of this issue was provided by the authors of the case study of the Hagmolenbeek River (NL) and is presented in Box 8.1.

According to the final HMWB and AWB guidance document, adverse effects on the specified uses are losses of/in important services (e.g. flood protection, recreation or navigation) or production losses (e.g. hydropower or agricultural goods). The guidance document also recognised that in assessing "significant adverse effects" on the specified uses, economic effects will play an important role (WFD CIS Guidance Docu-

Box 8.1. Consideration of the Costs of Restoration Measures

"*The fact that the costs of restoration measures* [of designation test 4(3)(a)] *are not included in the disproportionate costs* [of designation test 4(3)(b)] *can lead to the following situation:*

On the basis of the costs of other means [of designation test 4(3)(b)], *a water has been designated as natural or HMWB with a certain GEP. However, later in the River Basin Management Plan (RBMP) the costs of the restoration measures are considered too high and the water is designated as natural or HMWB with 'less stringent objective'. This forest of different objectives makes the application of the WFD very complex. Therefore, it is suggested that the 'less stringent objective' is replaced with the possibility of having – for instance as a result of the considerations in the RBMP – a longer time period to achieve the GEP or GES. This means that the objectives remain the same (as determined in the designation procedure), but restoration measures that are too expensive can be carried out at a later period (as a result of the RBMP).*"

Source: Case study on the Hagmolenbeek River, NL.

ment No. 4 2003). Therefore, the economic impacts of the restoration measures can be considered but not the costs of carrying out the measures themselves or the ability of the user to pay. Regarding the costs of the restoration measures themselves (including whether they are disproportionate or not), the final guidance document (Sect. 5.4.2 of the WFD CIS Guidance Document No. 4 2003) mentions, "*In assessing whether the restoration measures have 'significant adverse effects' [...], disproportionate costs cannot be used as an additional consideration beyond the assessment of significant adverse effects on the specified use*" (WFD CIS Guidance Document No. 4 2003). The costs of the restoration measures of the designation test 4(3)(a) should actually be assessed within the River Basin Management Plan.

Regarding the term "significance" of adverse effects, in general, the final guidance document on HMWB and AWB considered it not possible to derive a standard definition for "significant" adverse effect. "Significance" will vary between sectors and will be influenced by the socio-economic priorities of Member States (WFD CIS Guidance Document No. 4 2003).

According to the final guidance, however, it is possible to give an indication of the difference between "*significant* adverse effect" and "adverse effect". A significant adverse effect on the specified use should not be small or unnoticeable but should make a notable difference to the use. For example, an effect should not normally be considered significant, where the effect on the specified use is smaller than the normal short-term variability in performance (e.g. output per kilowatt hour, level of flood protection, quantity of drinking water provided). However, the effect would clearly be significant if it compromised the long-term viability of the specified use by significantly reducing its performance. It is important to undertake this assessment at the appropriate scale. Effects can be determined at the level of a water body, a group of water bodies, a region, a River Basin District, or on a national scale. The appropriate scale will vary according to the situation and the type of specified use or sector. It will depend on the key spatial characteristics of the adverse effects. In some cases, it may be appropriate to consider effects on more than one scale in order to ensure the most appropriate assessment. The starting point will usually be the assessment of local effects (WFD CIS Guidance Document No. 4 2003).

Significant Adverse Effects upon the Wider Environment

The authors of the UK-EW guideline report (Booker et al. 2002a) considered that the "wider environment" can possibly include archaeology, heritage, air quality, waste, energy, and townscape. In the case study on the Sarre River (F), the concept of "environment" was also considered in its wider sense and therefore included sites designated for their heritage or cultural interest.

Significant adverse effects upon the wider environment were reported in approximately one-third of the HMWB case studies. Adverse effects mainly related to flooding of inhabited areas and floodplains, the rise of the groundwater level or loss of wetlands. In the case of the Ruhr (D), the abandonment of the impounded lakes would decrease the groundwater level and would cause oxbow lakes to dry up. In the case of the Veluwerandmeren (NL), a more natural water level management would lead to the necessity of changing the water level management of Lake Issjel. As a consequence, water level management would have to be adapted in more than 30% of the Netherlands as well. The UK case studies on the Tummel and the Dee concluded that the proposed restoration measures would lead to the loss of features valued for nature conservation such as downstream wetland nature reserves.

The final guidance document on HMWB and AWB points out that the intent of assessing adverse effects upon the wider environment is to ensure that restoration measures required to achieve GES do not deliver environmental improvements for the water body whilst creating environmental problems elsewhere (Step 7.3 described in Sect. 5.4.3. of the WFD CIS Guidance Document No. 4 2003). The final guidance endorsed the approach of the HMWB case studies and concluded that a restricted definition of the environment would not be appropriate. Therefore, the environment is considered to include the natural environment and the human environment including archaeology, heritage, landscape, and geomorphology. In general, a significant adverse effect on the wider environment would exist if the damage to the wider environment caused by restoration measures exceeds the benefits for the improved water status itself (such as significantly increased CO_2 emissions or the generation and disposal of large quantities of demolition waste) (WFD CIS Guidance Document No. 4 2003).

8.3 Designation Test 4(3)(b): Assessment of Other Environmental Options

The second step of the designation process is the designation test 4(3)(b) (see Fig. 8.1). This test identifies whether the beneficial objectives served by the (artificial or) modified characteristic of the water body can reasonably be achieved by other means that are technically feasible, significantly better environmental options and not disproportionately costly.

8.3.1 Identifying "Other Means"

"Other means" (also referred to as alternatives or alternative options) were identified in all case studies except for the Wienfluss (A), Danube (A) and Mulde (D) (see Table 14.23 in Annex II). In the case on the Great Ouse (UK) (sub-study Wissey), Loosdrecht (NL) and Veluwerandmeren (NL), no hypothetical other means for flood protection were found. The interpretation of "other means" differed widely, and three main categories of "other means" can be summarised:

a *Replacement of the existing use:* Among others, this included replacing hydropower with other energy sources (e.g. Kemijärvi (SF), Danube (A), Beiarn (NO)), (partially) replacing navigation with rail or road transport (Great Ouse (UK), Veluwerandmeren (NL), Elbe (D)), and delivering water supplies from groundwater instead of surface waters (Loosdrecht, NL).

b *Displacement of the existing use to other water bodies/catchments:* This included displacement of recreational facilities (Veluwerandmeren (NL), Loosdrecht (NL)) and displacement of agricultural production, e.g. by reactivation of abandoned agricultural land outside the floodplain (Ruhr (D). This category also included shifting the water supply to other catchments (Lozoya (E)).

c *Maintenance of the existing use (with reduction of its environmental impact):* In the case of hydropower/water supply, this included the use of compensation flows and ecological water regulation regime (Lozoya (E), Suldalslagen (NO), Tummel and Dee (UK). In the case of navigation, this included the addition of non-navigable side channels as habitats (Great Ouse (UK)). In the case of agriculture, use of biological farming was considered (Hagmolenbeek, NL) or decreasing agricultural activity directly bordering the stream (Hagmolenbeek (NL)). For recreation, limitation of activities at certain locations and times was recommended (Hagmolenbeek, NL).

"Other means" falling under the categories "replacement of the existing use" as well as "maintenance of the existing use" often included measures that were already recommended as "restoration measures to achieve GES" (for test 4(3)(a)).[3] In these cases, it should be noted that for the purpose of defining the "other means", the "necessary restoration measures to achieve GES" were combined with additional measures in order to serve the beneficial objective of the existing use.[4]

It can be concluded that in the HMWB case studies, the choice of measures for the two tests (4(3)(a) and 4(3)(b)) appeared to be flexible and partly overlapping. A measure considered as a "restoration measure to achieve GES" was in some cases part of an "other means (which serves the beneficial objective)". However, based also on the agreed final guidance document on HMWB and AWB, the authors of this book consider that the use of the same measures in both designation tests could make subsequent cost evaluation complex because:

- Cost evaluation with regard to restoration measures (test 4(3)(a)) should be part of the River Basin Management Plan;

[3] Example: The removal of the estuary dam in the case of the Haringvliet Estuary (NL) was recommended both as a "restoration measure" as well as an "other means".

[4] In the case of the Haringvliet Estuary (NL), the "beneficial objective" was the supply of freshwater (at present possible due to the presence of the dam and lack of salt water intrusion). Therefore, the removal of the dam (other means and restoration measure) should be combined with additional measures to mitigate water salinisation so that the beneficial objective can still be served.

- Cost evaluation with regard to "other means" (test 4(3)(b)) follows the criterion of "disproportionate costs" and is relevant to the decision on the final designation as HMWB.

However, according to the final HMWB and AWB guidance document (Sect. 5.5.1 of the WFD CIS Guidance Document No. 4 2003) in considering the designation test 4(3)(b), it is important to distinguish between "restoration measures", covered under the designation test 4(3)(a), and "other means", which will deliver the beneficial objectives of the modified characteristics of the water body. Test 4(3)(b) should only consider the potential for "other means" and not "restoration measures" (WFD CIS Guidance Document No. 4 2003).

According to the final guidance document, other means may include the following options: replacement of the existing specified use with an alternative option (category *(a)* identified in the HMWB case studies) and displacement of the specified use to another water body to deliver the beneficial objectives (category *(b)* identified in the HMWB case studies). The third category (category *(c)*) identified in the HMWB case studies involving maintenance of the specified use is not in agreement with the final guidance document.

Finally, the guidance document noted that the partial replacement or displacement of the beneficial objectives of the existing use should also be considered, while not necessarily allowing the achievement of GES (WFD CIS Guidance Document No. 4 2003).

8.3.2 Assessment of "Technical Feasibility"

In case "other means" to deliver the beneficial objectives can be identified, it has to be first assessed whether these "other means" are technically feasible. In the HMWB case studies, approximately one-third of the proposed "other means" were considered to be either not technically feasible or questionable (see Table 14.23 in Annex II). The following aspects were reported as relevant in the HMWB case studies for the assessment of "technical feasibility":

- *Physical/engineering aspects such as:*
  - Substitution of groundwater supply by surface water, which was considered as impossible due to water pollution/salinity (Loosdrecht (NL)).
  - Lack of suitability of other power sources for short term peak production (Kemijärvi (SF), Tummel (UK), Dee (UK).
  - Lack of alternative crops for high groundwater levels (Hagmolenbeek (NL)).

  The UK case studies stated that engineering/physical factors should be the primary consideration for technical feasibility of other means. The Lower Wissey in the case study on the Great Ouse (UK) actually provided a clear example of a situation where there are no technically feasible alternatives to flood defence and embankments being available. The river is higher than the surrounding land and the only option would be to raise the surrounding land. However, this would require very large quantities of material across a very large area and would involve effectively burying high quality agricultural land.

- *Increasing needs for the beneficial objective:*
  - Increasing energy consumption in the future (Beiarn, Suldalslagen (NO));
  - Phasing out of nuclear power reduced the possibility of reducing hydropower (Umealven (S));
  - Increasing need for mineral extraction (Veluwerandmeren (NL)).
- *Feasibility within the given time frame:*
  - A fifteen-year time period for the realisation of the measures was assumed in the Haringvliet Estuary (NL) case study. The feasibility to demolish and rebuild a complete new dam within this time frame was questionable.
  - In the case study on the River Hagmolenbeek (NL), it was argued that catchments and streams, which had been adapted to agricultural needs during centuries, could not be restored to their natural condition within fifteen years.
- *Legal feasibility:*
  - Incompatibility with present operation license (Kemijarvi (SF)), long-term concessions (Veluwerandmeren (NL)), land use policy (Hagmolenbeek (NL)) or strategic objectives for the river basin (Dender (B)).
- *Social feasibility or acceptance:*
  - Acceptance and possible implementation, especially when it is based on voluntary co-operation (see extensive discussion in Box 8.2). In the case of the Dender River (B), it was proposed to change agricultural crops or switch to forestry; it was suggested that a social analysis might be needed since farmers were insecure about such changes.

    It could be argued that obtaining the agreement of affected stakeholders may affect feasibility. According to the UK case studies, the position of affected stakeholders may be a reason for derogation, but not designation. The UK case studies pointed out that it may not be appropriate to eliminate a measure just because agreement may be hard to reach. In the case study on the Seefelder Aach (D),

Box 8.2. Technical Versus Societal Feasibility, The Netherlands

It is difficult to determine the (technical) feasibility of certain measures and alternatives. Catchments and streams, which have been adapted to agricultural needs during centuries, cannot be restored to their natural condition in fifteen years. Furthermore, the implementation of most of the measures depends at the moment on voluntary co-operation and acceptability by the local public and the farmers. How this co-operation and acceptability will develop in the coming years is uncertain. In the Netherlands, the expropriation of land is presently only possible to reduce the risk of floods, but not to improve the ecological quality of waters. Next to these local aspects, the global, European and national agricultural market and politics determine the possibility for the farmers to compensate yield reduction as a result of environmental measures (e.g. higher groundwater levels, using less pesticides and nutrients), higher prices (e.g. more expensive products from biological farming) or obtaining subsidies (e.g. stimulation of biological farming).

At present, the agricultural sector is in a difficult economic position, and probably a lot of space will become available because farmers cannot survive financially. This space will be used by the remaining farmers as well as for other functions (nature development, housing). On the other hand, the difficult economic position will probably reduce the willingness of the farmers to cooperate in ecological restoration projects, as the nature lobby is not considered to be a partner.

Source: Case study on the Hagmolenbeek River, NL.

it was proposed to modify or abandon current land use structures, making the assumption that land owners were willing to sell their properties.

The final HMWB and AWB guidance document concluded that "technical feasibility" considerations include the practical, technical and engineering aspects of implementing the "other means". It addresses the question of whether "other means" of delivering the beneficial objectives of an existing specified use exist. There may be some circumstances where it is appropriate to consider social issues which constrain the development of "other means". The use of such social constraints should be fully explained within the RBMP (WFD CIS Guidance Document No. 4 2003).

8.3.3 Assessment of "Significantly Better Environmental Option"

In case an "other means" is found to be "technically feasible", it has then to be assessed whether it is a "significantly better environmental option". In the HMWB case studies, the assessment of "significantly better environmental options" was usually made on a qualitative basis on the local and the regional level. The following results can be summarised (Table 14.23 see Annex II):

- Better environmental options were mostly "other means" involving the maintenance of the existing use while reducing its environmental impact. The authors of this book would like to note again that this type of "other means", i.e. maintenance of the use, is not in agreement with the final guidance document on HMWB and AWB.
- The replacement of hydropower with energy production from nuclear or fossil fuels was not considered to be a better environmental option (due to increased air emissions, CO_2 production) (Kemijärvi (SF), Suldalslagen (NO), Emån (S), Dalälven (S), Umealven (S), Tummel (UK), and Dee (UK)). Alternative energy forms such as wind power, solar energy, geothermal energy as well as potential for energy savings were rarely discussed (except for the case study on the Umealven River (S) and the Suldalslagen River (NO)).
- In the case of hydropower, the assessment of "other environmental options" was considered very complex in the HMWB case studies, because hydropower could be replaced with a wide range of sources: New hydropower (usually less environmentally friendly), thermal (CO_2 emissions), nuclear power (radiation hazards), wind, solar and bio-energy. In most cases, reductions in hydropower production would reduce the value of other energy sources, since hydropower was used for balancing loads – from base load sources (thermal, nuclear) and from unpredictable sources (wind, solar). From the evaluation of the case studies, it can be concluded that the economic and environmental impacts of the replacement of hydropower were very complex to assess.
- The replacement of navigation with road/rail traffic was not considered a better environmental option (due to increased air emissions and extra traffic). For instance, this was the conclusion in the case study on the Dender River (B) by comparing external costs and environmental effects of different means of transport. An exception was the case study on the Elbe River (D), where a replacement of inland navigation with rail transport was regarded as a better environmental option; the con-

clusion was based on a comparison of ecological effects of the different means of transport referring to the emissions of air and noise pollution.

- In the case of the Great Ouse (UK), it was noted that the intended use of navigation requires a minimum water level that could only be achieved through the use of structures such as locks, weirs and sluices. Similarly, the channel needed to be wide enough to cope with the volume of boat traffic. This meant that the other means had to focus on ways to improve the ecological status of the river without affecting navigation, and consequently, the technically feasible "other means" involved mitigation works on by-pass and side channels rather than the main river.
- The abandonment or modification of current land use structures in the Seefelder Aach (D) was considered a better environmental option. Changes in land use patterns would result in positive environmental effects, while the restoration of the floodplains would positively affect species and habitats and reduce nutrient input through buffer strips.
- The displacement of the existing use to another water body or catchment was not considered as a significantly better environmental option, since it would result, only, in a displacement of the impacts (Lozoya (E), Loosdrecht (NL), Veluwerandmeren (NL)).
- Regarding flood protection, it seemed difficult to find alternatives. Construction of new flood levees, channels or storage reservoirs were not considered to be better environmental options (Great Ouse (UK), Kemijärvi (SF)).

It remained unresolved how to deal with possible aggregation effects. According to the authors of the case study on the River Suldalslagen (NO), an aggregation effect occurs when similar restoration measures are implemented throughout the country (e.g. increase of compensation flows and thereby loss of hydropower production). This might lead to the need for new energy production facilities and therefore raised again the question of technical feasibility and environmental consequences. Thus, assessing the resulting overall environmental "balance account" of such adjustments was extremely complex.

The final guidance document on HMWB and AWB suggested that when assessing other means as better environmental options, the following issues should be considered (WFD CIS Guidance Document No. 4 2003):

- *Scope of "environment" in better environmental option:* In order to ensure a consistent approach with the Article 4(3)(a) test of the WFD, the assessment should include – where appropriate – consideration of the "wider environment" such as archaeology and urban and other landscapes.
- *Issue of scale:* There is a range of scales at which the question of "better environmental options" can be assessed: local, regional, River Basin District, national, or international level. Clearly it may be appropriate to consider the impacts and benefits just on the water environment or on the wider environment (water, land, air). In the first instance it is suggested that the assessment should focus on local options. Further considerations should then be considered where appropriate.

According to the final guidance, it is clear that the most appropriate scale used to assess a "better environmental option" will depend on the kind of "other means" under consideration. Where there is uncertainty about the appropriate scale, an assessment should be carried out at different scales.

8.3.4
Assessment of "Disproportionate Costs"

"Other means" that are considered to be "technically feasible" and represent a "significantly better environmental option" should be subject to an assessment of whether they are "disproportionately costly". For assessing whether "other means" are disproportionately costly or not, the support material (guidance papers) agreed by the HMWB WG for conducting the case studies suggested the following three methods:

1. *Comparison of costs of alternatives*: This approach compares the existing costs of delivering the use, service or beneficial objective with the costs of "other means". Thus, for the existing modification, operation, maintenance costs as well as replacement cost should be considered while for the alternative the capital costs, operation, maintenance costs as well as possible foregone benefits have to be assessed.
2. *Comparison of overall costs and benefits:* This approach compares the overall costs and benefits of the existing modification and of the alternative. This is an assessment, which ensures that the modification provides an overall net benefit to society.
3. *Costs versus ability to pay*: This method assesses costs of alternatives with the ability to pay. Although ability to pay is not directly a designation process issue, it can be a useful way to assess different alternatives serving the same beneficial objectives.

These methods have been revised in the final guidance document on HMWB and AWB (WFD CIS Guidance Document No. 4 2003) and are summarised in the end of the following subsection on methods.

Methods for the Assessment of Disproportionate Costs

Several case studies did not carry out an economic analysis for the evaluation of disproportionate costs and based decisions mostly on qualitative estimations. Often, a full economic analysis was not possible due to the lack of data and because the costs of alternatives were often unknown and not possible to calculate given the time and resource constraints of the case studies. In some cases, it was also mentioned that a qualitative economic analysis was carried out instead of quantitative, because the case study authors were not sure of the extent of the economic analysis necessary for the purpose of HMWB designation.

Approximately half of the HMWB case studies carried out a relatively detailed economic analysis for the assessment of the proportionality of costs. The approaches used differed widely (Table 14.24 see Annex II).

One case study (Hagmolenbeek (NL)) followed the approach of "comparison of costs of alternatives" (method 1) and thereby compared the costs of the existing use and of the alternative. No case study used the approach "costs versus the ability to pay" (method 3). Method 2, "comparison of the overall costs and benefits", was followed in ten case studies; it was not, however, followed strictly due to the lack of either a cost-benefit analysis of the existing use or of quantification of the benefits of the alternative.

The methods used for the assessment of the proportionality of costs in the HMWB case studies can therefore be summarised as follows:

a Comparison of costs of alternatives (Hagmolenbeek (NL)).

b Comparison of costs and benefits of the alternative: if the resulting net-benefit was negative, the alternative was usually regarded as disproportionately costly (e.g. Ruhr (D), Dee (UK)).

In the case study on the River Ruhr (D), the calculation of costs included costs to replace existing uses, costs for the establishment of alternatives to achieve the same benefit according to Art. 4(3)(b) of the WFD, and real costs to realise measures. The costs were compared with the willingness to pay for nature protection (non use value) including the expected human benefit (use values). Costs were considered disproportionate when the "benefit-cost-quotient" was less than one (1), i.e. estimated costs were higher than the benefits.

c Comparison of benefits of the existing use and costs/benefits (including foregone benefits) of the alternative (cases of Suldalslagen (NO), Beiarn (NO) and Emån (S)). In this case, the following were calculated: benefits of the existing use such as turnover from power production, costs of the alternative (e.g. reduction in power production) and benefits of the alternative (e.g. increase of income from recreation). The outcome was based on the comparison of the existing use and the alternative in terms of production, income and turnover. The benefits of the existing use (hydropower in all three cases) were expressed in annual turnover from power production.

d Comparison of costs of the existing use and costs/benefits of the alternative (UK-EW case studies, i.e. Kennet, Tame, Great Ouse, Sankey Brook as well as the UK-NI case study on the Lagan River). In this case, detailed assessment *pro forma* tables were used; decisions were based on the calculation of costs (operation and maintenance) of the existing use, costs of the alternative (capital, operation and maintenance) and potential environmental benefits. The data were used for a benefit/cost transfer assessment.

e Absolute costs of the alternative (Dender (B), Elbe (D), Seefelder Aach (D), Lahn (D), Haringvliet Estuary (NL), Forth Estuary (UK), Tummel (UK), and Dee (UK)). In three German cases (Elbe, Seefelder Aach and Lahn), costs were assessed as non-disproportionate. In this case, decisions were based on the absolute costs of realising the alternative based largely on expert judgment. Different types of costs have been considered. For instance, in the case of Seefelder Aach, the costs considered for modifying or abandoning agriculture were costs for land purchase and for restoration measures, investment and implementation costs, operation and maintenance costs, and opportunity costs associated with financial losses of farmers due to lower yields. In the case of shutting down hydropower plants in the Tummel (UK), costs considered were the costs for the decommissioning of the dams and for the construction of fossil fuel power plants.

f Qualitative assessment was the basis for the decision on disproportionate costs in some case studies (e.g. Kemijärvi (SF), Rhone (F), Sarre (F), Veluwerandmeren (NL), and Umealven (S)).

The level of assessment also affected decisions on the disproportionality of costs. In the case of small hydropower plants in the study on the Lahn River (D), the level considered for "other means" (local or national) was discussed. According to the case study authors, both levels should be taken into account for the HMWB designation. On the local level, small hydropower plants were not considered profitable for electricity generation. On the national level, they were considered a rather expensive way to avoid CO_2 emissions.

Regarding the assessment of disproportionate costs of other means, the final guidance document on HMWB and AWB emphasised the focus on financial/economic costs. However, there may be some circumstances where it may be appropriate to consider social issues as part of the assessment of the disproportionality of costs (WFD CIS Guidance Document No. 4 2003).

According to the final guidance, in undertaking this assessment it is important to take account of likely or planned capital expenditures associated with the existing specified use; this should include planned expenditures up to 2027, where appropriate. This is particularly appropriate (and important) in cases where the existing specified use is associated with large-scale engineering works which are subject to regular maintenance, replacement or upgrading. This represents a key baseline, against which the incremental costs and benefits of the alternative "other means" are to be analysed and presented (WFD CIS Guidance Document No. 4 2003).

The final guidance recommended the following two options as methods for assessing the proportionality of costs (Sect. 5.5.4 of the WFD CIS Guidance Document No. 4 2003):

a Comparison of cost alternatives: Disproportionate costs can be determined by assessing the incremental costs and environmental impacts of the "other means". The benefits of the existing specified use and the alternative are assumed to be the same. The main cost elements to be considered are:
  - For the existing situation: operational and maintenance costs, and capital costs for necessary replacements (including investment and interest costs);
  - For each option/alternative ("other means"): capital costs (including investment and interest costs), operational and maintenance costs, and possible foregone benefits from changes in economic activities (e.g. reduction in agricultural production resulting from the development of a retention area as an alternative to dikes for preventing floods).

b Comparison of overall costs and benefits: Disproportionate costs can be determined by comparing the overall costs and benefits of the existing modification and the alternative ("other means"). In this assessment, the overall net benefit to society of the modification and the alternative are compared. The main elements that are to be considered include:
  - Costs as listed in *(a)*;
  - Benefits of the existing specified use; and
  - Benefits of the alternative, especially benefits gained from higher ecological status (e.g. angling, recreation).

Calculation of Costs and Benefits

In general, no principal methodological problems were mentioned to obtain figures for costs of the alternatives. It was however noted that a detailed cost evaluation was very time-consuming and was not appropriate within the timeframe of the case studies.

Some problems appeared when attempting to evaluate the expected benefits of the alternatives. Benefits of the alternative included mainly benefits gained from higher ecological status and benefits to the ecology (e.g. increased biodiversity). In this respect, the authors of several case studies reported that there were no robust and agreed methods to assess benefits to the ecology such as enhanced biodiversity and beauty of the landscape. The most common and practical approach (e.g. in Kemijärvi (SF), Suldalslagen (NO), Emån (S), Great Ouse (UK), Tummel (UK), and Dee (UK)), was to estimate the economic benefits gained from higher ecological status on the basis of fishing/angling. In the case studies on the rivers Dee and Tummel (UK), the welfare measures of salmon fishing were predicted by the change in salmon population in the affected areas. The benefits were estimated by current rents for privately owned beats based on an average value per km. The Swedish case study on the Emån River estimated the economic value of fish stocks by assessing the willingness to pay for recreational fishing. In several case studies such as the Seefelder Aach (D), the Haringvliet Estuary (NL), Forth Estuary (UK), Tummel (UK), and Dee (UK), it was not possible to quantify the benefits from higher ecological status, since the effects were very complex and no studies on monetary values were available.

It was also proposed that recreational factors should be considered when determining environmental benefits. When, in specific, determining environmental benefits to bird populations as well as fisheries, the impact on overall populations should be considered.[5]

With regard to the pressure of "agricultural land take", the availability of generic estimates of the benefits (and costs) of managed retreat could be valuable for the HMWB designation process (as in the case on the Forth Estuary, UK). In the latter case, it was difficult to decide how to deal with the lost agricultural output, should land take be reversed.

In the case study on the River Dee (UK), it was noted that wider issues of greater species diversity, or abundance measures were largely unquantified. According to the authors of this case study, there was a lack of existing suitable studies to place adequate monetary values on these kinds of ecological improvements. Thus, the development and recognition of appropriate methodologies for quantifying the wider benefits of enhancing water body ecology should be a priority. It was proposed that the use of the SERCON methodology,[6] which quantifies and allows comparisons of conservation values, should be considered.

A cost-benefit analysis was often proposed as the best method to meet the requirements of the HMWB designation process because it ensures that those other means are chosen which provide a net benefit to society. It was however made obvious that the assessment of benefits is often difficult. As an alternative, the case study on the Seefelder Aach (D) proposed the method of benefit transfer as a suitable and cost-

[5] From the material agreed during the meetings of the case study subgroup on hydropower.
[6] System for Evaluating Rivers for Conservation (SERCON) is a method for defining river conservation value in which river features and the impacts on river systems are assigned scores. Information on each attribute is used to create a picture of a river in terms of traditional conservation criteria such as Naturalness, Species Richness, and Rarity (Boon et al. 2002).

effective approach, which uses the monetary values estimated at one study site and transposes them to another site. In the UK-EW case studies, the method of benefit transfer was also applied. The authors of the UK-EW guideline report discussed the use of benefit transfer and pointed out that its use for the assessment of environmental benefits (e.g. fishery value and recreation) is not standard practice. They also argued that a key issue that always arises with the use of benefits transfer is the validity of the transfer exercise and whether it will result in an under- or over-estimate of benefits. This is a particular concern where the original valuation was carried out in isolation from the national and regional context of the general environmental quality or availability of such sites.

Meaning of Disproportionality

Disproportionality does not mean that costs of measures simply exceed quantifiable benefits. An evaluation table for disproportionate costs, which also takes qualitative benefits into account to avoid a bias towards "quantifiable" benefits, was provided in the UK-EW guideline report (Booker et al. 2002a) (Table 8.4). This table was accompanied by a sensitivity analysis and was part of the UK-EW guideline report on the identification, assessment and designation of rivers, providing a scoping methodology for the identification and designation of HMWB under the WFD. In the UK-EW guideline report, the reader can find a series of *pro formas* to work through the proposed methodology for the HMWB designation tests. This includes detailed assessment of the costs and benefits of the existing use and the proposed "other means".

The final part of the detailed assessment summarised the findings and undertook a limited sensitivity analysis to check the robustness of the assessment. The first step was to make sure that all costs and benefits were given in present value terms. This was done by discounting all of the costs and benefits of the existing use described in the same way that the capital and operating costs of the "other means" were discounted. Once all of the costs and benefits had been converted to present value costs, they could be compared.

According to the authors of the UK-EW guideline report, most economic methodologies consider an option to be "worthwhile" if the benefits outweigh the costs (Booker et al. 2002a). Two different ways of presenting this comparison are usually used:

- Net present value – this is simply the benefits minus the costs; and
- Benefit-cost ratio – this is the benefits divided by the costs.

When the net present value (NPV) is greater than zero and the benefit-cost ratio (BCR) is greater than one, the option is considered economically "worthwhile". However, there may be some uncertainty within both the cost and benefit estimates. In the UK-EW case studies, this uncertainty was tested through the proposed sensitivity analysis. In the sensitivity analysis, it was possible to examine changes in any of the values used in order to assess the impact on the final benefit or cost estimates. However, it was most useful to consider where the estimates were most uncertain. For example, if one had assumed that the measure would result in an improvement from a poor quality coarse fishery to a high quality coarse fishery (an increase in value of

£13 230 km^{-1} year^{-1}), one might want to assess how the benefits would change if the fishery only improved to a "good quality coarse fishery" (an increase in value of £4 230 from a poor quality coarse fishery).

It was useful to consider what (and how many) changes were necessary for the costs to be higher than the benefits (if benefits were originally greater than costs) or for the benefits to be higher than costs (if costs were originally larger than benefits). Then it was decided whether the "new" assessment looked reasonable or not, i.e. whether it was necessary to make unreasonable assumptions in order to make the measure

Table 8.4. Evaluation of disproportionate costs, UK-EW. *Source:* UK-EW guideline report (Booker et al. 2002a)

Costs versus benefits	Disproportionately costly?
Costs outweigh benefits significantly (>2:1); no significant qualitative benefits	Yes
Costs outweigh benefits slightly (sensitivity analysis shows benefits could outweigh costs but only if unreasonable changes to assumptions are made); no significant qualitative benefits	Yes
Costs outweigh benefits slightly (sensitivity analysis shows benefits could outweigh costs); significant qualitative benefits (particularly environmental)	Unsure – qualitative benefits may mean measure is "worthwhile"
Costs outweigh benefits slightly (sensitivity analysis shows benefits could outweigh costs if reasonable changes to assumptions are made); no significant qualitative benefits	Unsure – can the costs be reduced (or benefits increased) by combing the measure with other measures; or are there other measures with larger benefit-cost ratios?
Costs outweigh benefits slightly (sensitivity analysis shows benefits could outweigh costs if reasonable changes to assumptions are made); significant qualitative benefits (particularly environmental)	Unsure – can the costs be reduced (or benefits increased) by combing the measure with other measures; or are there other measures with larger benefit-cost ratios? Qualitative benefits may make the measure "worthwhile"
Benefits and costs are very similar; no significant qualitative benefits	Unsure – can the costs be reduced (or benefits increased) by combing the measure with other measures; or are there other measures with larger benefit-cost ratios?
Benefits and costs are very similar; significant qualitative benefits (particularly environmental)	Probably no
Benefits outweigh costs slightly; no significant qualitative benefits; (sensitivity analysis shows costs could outweigh benefits if reasonable changes to assumptions are made)	Unsure – can the costs be reduced (or benefits increased) by combing the measure with other measures; or are there other measures with larger benefit-cost ratios?
Benefits outweigh costs slightly; no significant qualitative benefits; (sensitivity analysis shows costs could outweigh benefits but only if unreasonable changes to assumptions are made)	Probably no
Benefits outweigh costs slightly; significant qualitative benefits (particularly environmental); (sensitivity analysis shows costs could outweigh benefits but only if unreasonable changes to assumptions are made)	No
Benefits outweigh costs significantly (>2:1)	No

"worthwhile" or "not worthwhile". Details of the sensitivity analysis were recorded in the final *pro forma* of the UK-EW guideline report. Once the sensitivity analysis was completed, it was decided whether the measure was considered "disproportionately costly" or not. Guidance to help make this decision was given in Table 8.4.

In the context of the final guidance document on HMWB and AWB, it was not possible to clearly define by how much the costs must exceed the benefits before they become disproportionate (WFD CIS Guidance Document No. 4 2003).

8.4 Examples of Different Designation Approaches

As mentioned in the introduction to this book, the HMWB case studies aimed to test twelve guidance papers (support material for conducting the case studies) produced by the HMWB working group prior to the publication of the CIS guidance document on HWMB and AWB. The twelve guidance papers also proposed a draft process on the identification and designation of HMWB and AWB (see Fig. 2.2, p. 15). Some case studies did not strictly follow the draft process of Fig. 2.2 (p. 15). These case studies instead applied similar alternatives. It should be pointed out that the alternative approaches presented below provided no strict separation of "restoration measures" (test 4(3)(a)) and "other means" (test 4(3)(b)), which is not in complete agreement with the CIS guidance document on HMWB and AWB. The latter requires the distinction of restoration measures and other means (Sect. 5.5.1 of the WFD CIS Guidance Document No. 4 2003).

The alternative process on provisional identification and designation of HMWB used in the German case studies was based on positive/negative lists of physical alterations with the aim to simplify the designation process. A positive list includes "non-significant" physical alterations, for which measures could be identified to achieve GES. A negative list included "significant" physical alterations, whose presence led to the provisional identification and designation of a water body as heavily modified (see Sect. 7.3 of this book). The approach adopted in the German case studies on the rivers Elbe, Lahn and Seefelder Aach is described in Box 8.3 and Fig. 8.2 (compare also with Fig. 7.2, p. 101, on provisional identification in the German case studies).

The alternative process described in the German case studies on the rivers Elbe, Lahn and Seefelder Aach has not been explicitly adopted in the final guidance document on HMWB and AWB. However, it can be considered a useful approach especially for the purpose of identifying significant anthropogenic pressures on water bodies.

The UK-EW also proposed an alternative scheme for HMWB designation as described in Box 8.4 and in Fig. 8.3. This entailed a pre-screening methodology to screen out water bodies with minor modifications. This approach was considered very useful by the UK-EW case study authors because of the large number of water bodies to be considered for the HMWB designation. The subsequent steps of the proposed designation scheme, after pre-screening, involved the screening step and the step of detailed assessment. A short description of the steps of the overall approach is given in Table 8.5.

As already mentioned in Chap. 5, a screening step similar to the UK approach has been included in the final guidance document on HMWB and AWB (WFD CIS Guidance Document No. 4 2003) (see Step 3 in Fig. 2.1, p. 12), This screening step aims to

Box 8.3. Designation Approach of the German Case Studies

According to Fig. 8.2, following the HMWB provisional identification (as described in Sect. 7.3 of this book), restoration measures were identified. In the context of an economic analysis, the restoration measures were examined regarding their possible significant adverse effects on specified uses. In case the measures impaired existing specified uses, "better environmental options" were proposed. The proposed measures were checked for their technical feasibility and financial proportionality. According to the results of the economic analysis, a so-called "2nd Negative/Positive list (of physical alterations and impacts)" was formulated, which led to the final designation as HMWB or natural water (as compared to the 1st Negative/Positive list, which led to the provisional identification as HMWB or natural). Physical alterations and impacts included in the 2nd Positive list did not lead to a designation of water bodies as heavily modified but to the application of the environmental objective "Good Ecological Status" (natural water). The 2nd Negative list led to designation as HMWB and the environmental objective was "Good Ecological Potential".

Source: Case study on the Seefelder Aach, D.

Box 8.4. Designation Approach of the UK-EW Case Studies

The UK-EW guideline report for HMWB designation used a series of *pro formas*. Completion of each *pro forma* gave a record showing why a particular water body had or had not been designated.

The overall process was divided into two stages: screening (Step 1) and detailed assessment (Step 2). Screening provided a rapid methodology to help decide whether the ecological quality of the river could be improved using measures that would not cause a "significant adverse effect" and were not "disproportionately costly". If one were unable to answer "yes" or "no" for each rehabilitation or alternative measure at the end of the screening stage, then the detailed assessment allowed for the collection and use of more data on which to base the decision. However, one should be careful when stating "yes" or "no", since the reason for this conclusion must be justifiable and must be given in the relevant *pro formas*.

The *pro formas* were split into two levels of assessment:

- A screening stage where those measures that would have clearly significant adverse effects on the intended uses that were not technically feasible or were clearly disproportionately costly were screened out. Such measures were assumed to fail either of the two tests, as given in Article 4(3)(a) (for significant adverse effects) and Article 4(3)(b) (for technical feasibility and cost disproportionality) of the WFD; followed by
- A detailed assessment methodology where those measures which had not been screened out were then considered in greater depth. This included the assessment of costs of the measures and comparison of these costs with a range of potential benefits. Thus, the more detailed part of the assessment took the form of a cost-benefit analysis. There was also provision for those impacts that could not be expressed in monetary terms to be given a qualitative "rating" of the relative size of impacts. This approach helped to ensure that the methodology was applied consistently and reduced the potential for a decision to be made on the quantified impacts only.

Source: UK-EW guideline report (Booker et al. 2002a).

screen out water bodies with only minor modifications and can be considered equivalent to the pre-screening step proposed by the authors of the UK-EW case studies.

In the case study on the River Sarre (F), an overall decision-making process for HMWB designation was proposed (see Fig. 7.1, p. 99). This consisted of a physical fil-

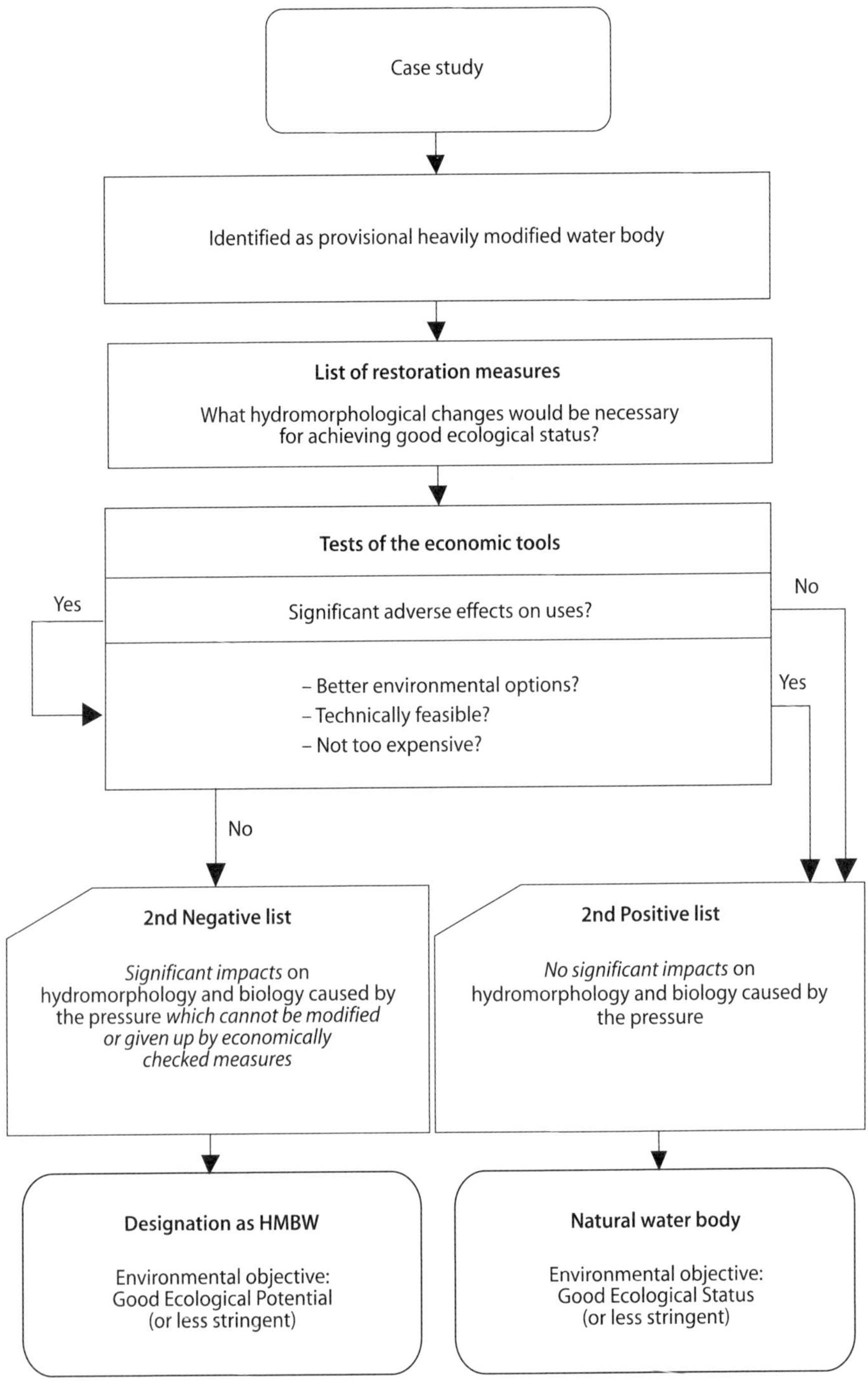

Fig. 8.2. Designation approach of the German case studies. *Source:* Case study on the Seefelder Aach, D

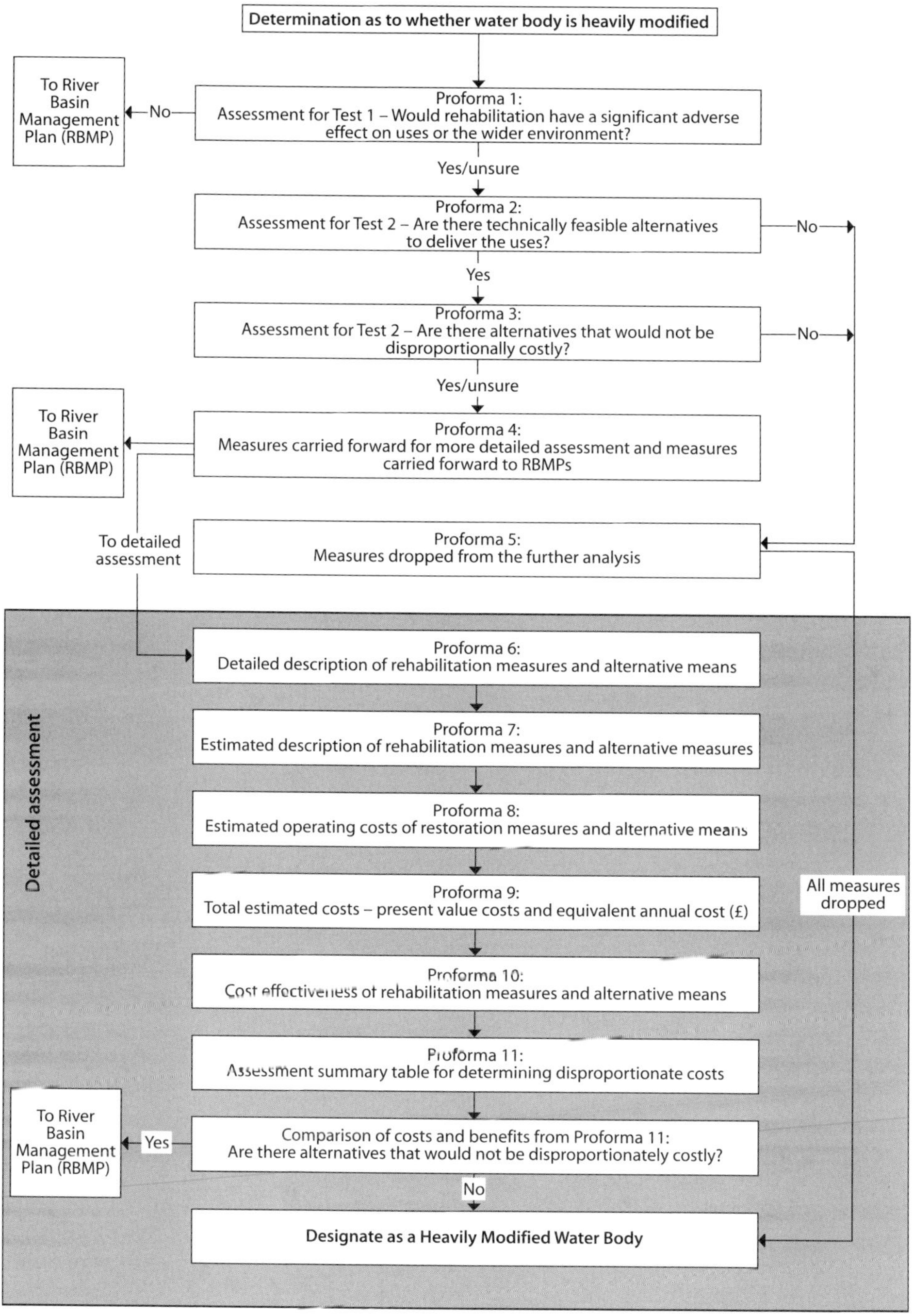

Fig. 8.3. Scheme of the designation approach of the UK-EW case studies. *Source:* UK-EW guideline report (Booker et al. 2002a)

Table 8.5. The four steps of the designation approach of the UK-EW case studies. *Source:* UK-EW guideline report (Booker et al. 2002a)

Stage	Decision results
Pre-screening, to screen out water bodies with minor modifications	Not HMWB, Potential HWMB or Borderline HMWB (proceed to screening)
Screening	HMWB or proceed to full assessment
Full assessment	HMWB or not HMWB or detailed investigation
Detailed investigation	HMWB or not HMWB

ter (question Q1) which led to the selection of precandidate water bodies. The physical filter was followed by the biological filter (question Q2), which led to the selection of candidate water bodies (provisional identification of HMWB). Questions Q3, Q4 and Q5 led to the designation as HMWB considering technical feasibility, adverse effects and cost disproportionality.

8.5 Overview of Results on HMWB Designation

An overview of the main results of the designation process is given in Table 14.21 (see Annex II). In thirteen case studies,[7] at least one of the provisionally identified HMWB was *not designated* as heavily modified. In some case studies such as the Bregenzerach (A), Great Ouse (UK) and Seefelder Aach (D), some restoration measures required to achieve GES[8] (designation test 4(3)(a)) were not considered to have any significant adverse effects on the specified use or the wider environment, and thus the respective water bodies were regarded as "natural". In some cases such as the Elbe (D) and the Lahn (D), the existing use[9] was regarded as low or non-profitable, and therefore the overall net-benefit of achieving Good Ecological Status was considered to be higher than the benefit of the existing use, thus not justifying the designation as HMWB. In the case of the Mulde River (D), one provisionally identified HMWB was not designated as HMWB, because according to the case study authors, failure to achieve Good Ecological Status resulted from pollution and acidification by mine drainage and not from physical alterations.

In nineteen case studies, all water bodies were designated as HMWB. The reasons for the designation of water bodies as heavily modified can be briefly summarised as follows:

With regard to designation test 4(3)(a): The majority of the restoration measures identified in the HMWB case studies were assessed to have a significant adverse effect on the specified uses and to a lesser extent also on the wider environment. For

[7] Bregenzerach (A), Elbe (D), Seefelder Aach (D), Lahn (D), Ruhr (D), Mulde (D), Sarre (F), Rhone (F), Nestos (GR), Great Ouse (UK), Forth Estuary (UK), Dee (UK), and Lagan (UK).

[8] Bregenzer Aach: tunnel for hydropeaking flow; Great Ouse: non-navigable side channels as habitats, Seefelder Aach: restrictions for agriculture.

[9] Elbe: Inland navigation, Lahn: relatively small hydropower plants.

some of the identified restoration measures, the significance or not of their adverse effects on the specified uses or the wider environment remained unclear (see Tables 14.21 and 14.22 in Annex II).

With regard to designation test 4(3)(b): In most cases, the "other means" identified to deliver the beneficial objectives of the artificial or modified characteristics of the water body were considered to be either not a better environmental option or disproportionately costly. In fewer cases, the identified "other means" were not considered further due to their technical infeasibility.

8.6 Conclusions and Discussion

The main conclusions on the process of designation of HMWB can be summarised as follows:

- An exact cause-effect-relationship between a restoration measure and the resulting ecological condition was very difficult to derive for most ecosystems and was accompanied by a high degree of uncertainty. This lack of knowledge raised questions with regard to the juridical and financial consequences of not achieving GES (or GEP) by 2015. The authors of the case study on the River Hagmolenbeek (NL) suggested that it may be more practicable to describe the WFD objectives also in hydromorphological and water quality terms in addition to biological terms.

  In cases where poor ecological status resulted from both physical alterations and low water quality, the effect of restoration measures on water quality and the ecological status was nearly impossible to differentiate with current methodologies and knowledge.
- The distinction between "restoration measures" (test 4(3)(a)) and "other means" (test 4(3)(b)) was not clear in the case studies. The choice of measures considered under the two tests (4(3)(a) and 4(3)(b)) was flexible. Measures considered as "restoration measures to achieve GES" were considered in some case studies as part of "other means" (which serve the same beneficial objective).

  However, according to the final guidance document on HMWB and AWB, it is important to distinguish between "restoration measures" and "other means". "Other means" may involve displacement of the specified use to another water body or replacement of the existing specified use with an alternative option to deliver the beneficial objectives (WFD CIS Guidance Document No. 4 2003). The third category of "other means" identified in the HMWB case studies that involved maintenance of the specified use (with reduction of its environmental impact) is not in agreement with the final guidance document.
- The need to apply quantitative or qualitative appraisal techniques regarding the designation tests 4(3)(a) and 4(3)(b) differed on a case-by-case basis. For "clear-cut situations", a qualitative description was sufficient for reaching a decision on HMWB designation. For instance, the main use of the reservoirs of the Lozoya River (E) was to provide 90% of the water supply of Madrid. In this case, it was straightforward that there was no better environmental option. However, this decision and the reasoning for it have to be transparent and documented for stakeholders. Another example was the Ruhr River (D), which involves a harbour area. The only way to achieve

GES would be to give up the harbour or to restrict harbour activities. As the harbour in question is the world's biggest inland harbour with a trading volume of more than two million tons per year, a detailed quantitative analysis of the impact of abandoning the harbour or of changing the harbour activities to achieve GES did not seem to be of any use, since there was no realistic chance that it would happen. However, once again this decision and the reasoning for it have to be transparent and documented.

- None of the case studies involved a consultative forum or an expert panel group to assess the significance of adverse effects upon the specified uses or the wider environment. According to the authors of the case study on the Hagmolenbeek River (NL), this may be due to the fact that in many cases the consideration of "significance" was subjective and political, and different stakeholders (experts) would probably have opposing views. Moreover, there was a need to clarify the type of experts that should be involved. Another reason could be the short timescale of the HMWB case studies and the fact that these consultative fora were likely to be set up at a later date for the RBMPs.
- Regarding cost-benefit analysis, it was not easy to relate costs with benefits to ecology. There was a need for methodologies on quantitative estimates of the benefits gained from the improvement of the ecological status. Due to the existing methodological difficulties, a systematic underestimation of these benefits occurs. This issue was common in many case studies and therefore pointed out the need for guidance on appropriate methods or values to promote consistency.
- The presence of "significant" physical alterations should not inevitably lead to the designation of a water body as heavily modified. This was exemplified in the German case studies on the Elbe, Lahn and Seefelder Aach, where none of the identified water bodies was designated as heavily modified, although they were affected by "significant" physical alterations and hydromorphological changes. In the German case studies, it was further concluded that the number of water bodies that are clearly designated as heavily modified should be the exception rather than the rule. According to the authors of these case studies, this may also be the case for other countries with comparable population densities and infrastructure.

  Nevertheless, this was not necessarily a commonly held view, particularly as it is clear that the WFD is concerned with sustainability and HMWB designation with the maintenance of justifiable human uses. In case of HMWB designation, it should be emphasised that designation is not a derogation and that the GEP objective is still a very stringent requirement. Member States will have to bring HMWB ecological status up to a condition that is close to the maximum possible given the essential modification that supports the use. Member States can derogate from GEP only if this can be fully justified on the basis of technical infeasibility or disproportionate costs.
- As mentioned in the beginning of this book, the HMWB case studies were carried out prior to the publication of the final CIS guidance document on HMWB and AWB. The HMWB identification and designation approach followed in the case studies was based on the draft process of Fig. 2.2 (p. 15) which was agreed by the CIS Working Group on HMWB during its initial activities. Some case studies did not strictly follow this proposed identification and designation process but presented similar al-

ternatives. These alternatives mainly aimed at simplifying the process as well as at concentrating effort on complex cases of designation. Some elements of these alternative approaches have been considered for the production of the final CIS guidance document on HMWB and AWB. For instance, the UK-EW guideline report (Booker et al. 2002a) proposed a pre-screening methodology to screen out obviously Heavily Modified Water Bodies from obviously natural water bodies. In the final guidance document on HMWB and AWB (WFD CIS Guidance Document No. 4 2003), a screening step has been included (see Step 3 in Fig. 2.1, p. 12), which can be considered equivalent to the pre-screening step of the UK-EW guideline report.

In the HMWB case studies, the need to clarify the meaning of the following terms was expressed:

- "*Significant*". As already mentioned earlier in this chapter, according to the final HMWB and AWB guidance document (Sect. 5.4.2 of the WFD CIS Guidance Document No. 4 2003), it was not considered possible to derive a standard definition for "significant" adverse effect. "Significance" will vary between sectors and will be influenced by the socio-economic priorities of Member States.
- "*Beneficial objectives*" (of Art. 4(3)(b)). It remained unresolved whether they cover only the beneficial objective of "specified uses" as in (ii–v) of Art. 4(3)(a), or also (indirect) beneficial objectives with regard to the wider environment as in (i) of Art. 4(3)(a) of the WFD.

  This was clarified in the glossary of the final HMWB and AWB guidance document (see Sect. 8.1 of the WFD CIS Guidance Document No. 4 2003) as follows: "Beneficial objectives are the benefits that result from the artificial or heavily modified characteristics of a water body and can include "specified use"-related or environmental benefits".
- "*Disproportionate costs*". In the case studies, there was a need for clearer guidance on the evaluation as well as the definition of "disproportionate costs". As already mentioned, the final HMWB and AWB guidance document (Sect. 5.5.4 of the WFD CIS Guidance Document No. 4 2003) considered that the assessment of disproportionate costs is likely to focus on financial/economic costs. The following two options were recommended for their evaluation: comparison of cost alternatives and comparison of overall costs and benefits. It was pointed out that it was not possible to define by how much the costs must exceed the benefits before they become disproportionate. In the context of economic assessments, the guidance document produced by the CIS Working Group WATECO should be considered (WFD CIS Guidance Document No. 1 2003).
- "*Technical feasibility*". In the case studies, there was also the need for further clarification on the scope of technical feasibility, especially in regard with time frame and social feasibility or acceptance. The final HMWB and AWB guidance document (Sect. 5.5.2 of the WFD CIS Guidance Document No. 4 2003) stated that technical feasibility considerations should include the practical, technical and engineering aspects of implementing the "other means". There may be some circumstances where it is appropriate to consider social issues that constrain the development of "other means".

The need for further discussion and clarification was identified on the following issues:

- The method for assessing the effectiveness of individual restoration measures (and the risk of failing GES) as well as of "packages" of restoration measures required to achieve GES;
- The method for determining the "business as usual" baseline (status quo) in the assessment of cost effectiveness and the proportionality of costs.

Chapter 9

Definition of Maximum Ecological Potential

9.1 Introduction

As indicated in the Heavily Modified and Artificial Water Bodies (HMWB and AWB) designation process, following the designation of HMWB and AWB, it is necessary to identify the appropriate reference conditions and environmental objectives for AWB and HMWB (Steps 10 and 11 of Fig. 2.1, p. 12). While for natural waters the reference condition is the High Ecological Status (HES), for HMWB and AWB it is the Maximum Ecological Potential (MEP) as defined in Step 10 of Fig. 2.1. According to the WFD, the Maximum Ecological Potential is defined as the state where "the values of the relevant biological quality elements reflect, as far as possible, those associated with the closest comparable surface water body type, given the physical conditions, which result from the heavily modified characteristics of the body" (WFD Annex V 1.2.5).

The hydromorphological conditions of an HMWB or AWB at MEP "are consistent with the only impacts on the surface water body being those resulting from the artificial or heavily modified characteristics of the water body once all mitigation measures have been taken to ensure the best approximation to the ecological continuum, in particular with respect to migration of fauna and approximate spawning and breeding grounds" (WFD Annex V 1.2.5).

Physicochemical and nutrient conditions as well as specific non-synthetic pollutants at MEP must "correspond totally or nearly totally to the undisturbed conditions associated with the surface water body type most closely comparable to the Artificial or Heavily Modified Water Body concerned" (WFD Annex V 1.2.5). The requirements for specific synthetic pollutants are the same as those for unmodified water bodies with "concentrations close to zero and at least below the limits of detection of the advanced analytical techniques in general use" (WFD Annex V 1.1.5)

For establishing the MEP "the quality elements applicable to Artificial or Heavily Modified Water Bodies shall be those applicable to whichever of the four natural surface water categories (river, lake, transitional water, coastal water) most closely resembles the Heavily Modified or Artificial Water Body concerned" (WFD Annex V 1.1.5).

The methods for establishing the values of the reference conditions (MEP) for HMWB are identified in Annex II (1.3) of the WFD and resemble the procedure for natural water bodies. Reference conditions may be:

i Spatially-based, i.e. established by description and classification of other surface waters suitable as reference sites;
ii Derived by hind-casting or predictive modelling, using historical, palaeo-ecological and other available data;

iii Based on a combination of *(i)* and *(ii)*; or
iv Established by expert judgment.

When the HMWB case studies were carried out, reference conditions had not yet been established either for natural or for HMWB. Therefore, a robust definition of MEP was impossible at that time. However, the "best-possible" approximation to the MEP definition was attempted in the HMWB case studies.

The support material (guidance papers) agreed by the HMWB WG for conducting the case studies included preliminary guidance on the issue of the definition of Maximum and Good Ecological Potential. However, many of the case studies followed their own approaches for setting Maximum and Good Ecological Potential. This chapter will hence focus on the relevant approaches and results of the case studies. This chapter will also present the approaches used in relation to some of the discussion points of the final guidance document on HMWB and AWB on the establishment of MEP, to the extent possible.

9.2 Methods for the Definition of MEP

Due to the complexity and uncertainty regarding the definition of reference conditions, it has proven difficult to summarise the methods and results on MEP. However, it could be argued that roughly two different approaches for the definition of MEP were followed in the HMWB case studies (for details on each case study see Table 14.26 in Annex II):

i Reference conditions for the relevant quality elements were the starting point. Based on these reference conditions, the necessary mitigation measures to achieve MEP were identified (all mitigation measures that have no significant adverse effect on the specified use). This approach was encountered among others in the Austrian case studies on Danube, Wienfluss and Bregenzerach, in the case studies on Lozoya (E), Loosdrecht (NL), Veluwerandmeren (NL), Kemijärvi (SF) as well as in the UK-EW case studies (on the latter, see Box 9.1).
ii Defining "all mitigation measures" (which have no significant adverse effect on the specified use) and predicting their ecological effect was considered as the starting point. The definition of MEP was thereby based on the predicted ecological effects of the mitigation measures. This approach was encountered in the case studies on Beiarn (NO) and Emån (S).

In summary, according to the first approach, the definition of MEP was based on reference conditions similarly to the definition of "high ecological status" for natural waters, while the second approach emphasised the mitigation measures. The first approach was followed by the majority of the HMWB case studies and is in line with the final guidance document on HMWB and AWB. The final guidance document also emphasised the importance of mitigation measures, which have been the focus of the second approach, as part of the definition of the hydromorphological conditions at MEP. Establishing these hydromorphological conditions is one of the first steps in defining MEP since it is these conditions that are impacted by the physical alterations and will, primarily, dictate the ecological potential of an HMWB or AWB.

Box 9.1. Definition of High Ecological Status and Maximum Ecological Potential, UK-EW

The UK-EW guidelines (Booker et al. 2002a) suggested a general methodology on the selection of reference sites for ecological potential (as well as ecological status). The methodology consisted of an integrated assessment for ecological status and ecological potential rather than two separate assessments. This was to ensure that strict time deadlines can be met. The general methodology was based on the comparison with reference conditions according to the river typology and was supplemented by expert opinion or backed by historical records. For the assessment of ecological potential, the water quality stretches had to be filtered by the modifications instead of the habitat quality score, as described by the River Habitat Survey (RHS).

Identifying reference sites involved matching several site-specific, map-based attributes from the RHS reference network for the definition of reference conditions. The steps involved included:

1. Select the potentially best reference sites by comparison of each of the RHS survey sites in each test section to the nearest 100 sites from the dynamic RHS classification system.
2. For Maximum Ecological Potential, match the RHS survey sites in the section under consideration with similar sites with:
   - A modification feature present, e.g. embanked;
   - A similar low score in the sub-division of Habitat Quality Assessment score (e.g. bank structure);
   - A similar habitat descriptor (e.g. water depth for plants).
3. Reduce potential reference sites to only three RHS sites by removal of those not similar in the following order:
   - Deep and surface geology;
   - Distance, location (i.e. another catchment or Environment Agency region);
   - Different physical characteristics (width, flow type, discharge category).

Although the normative definitions of ecological potential classifications found in the WFD represent a crude methodology, they provide clear qualitative classification of maximum, good and moderate. However, this type of approach gives freedom for interpretation and will produce discrepancies in the classification of ecological potential due to the variability found between different expert opinions. In the proposed methodology, the three qualitative categories (i.e. maximum, good and moderate) were divided into three quantitative categories based on deviation from reference assemblages.

The suggested bands for ecological potential (level of deviation from reference) were:

- Maximum Ecological Potential: 1%;
- Good Ecological Potential: 2–25%;
- Moderate Ecological Potential: 26–50%.

The same quality elements and potential scoring elements used for the classification of ecological status (see Table 6.3) were used for the definition of MEP and the classification of ecological potential.

Source: UK-EW guideline report (Booker et al. 2002a).

In some case studies such as the ones on Hagmolenbeek (NL), Tummel (UK-S) and Forth Estuary (UK-S), a combination of both approaches *(i)* and *(ii)* outlined in the beginning of this section was used for the definition of MEP. An example has been extracted here from the case study on the Forth Estuary (UK-S) (Fig. 9.1), where MEP was derived by two independent routes. The first route predicted the effects of eliminating non-physical pressures and applying appropriate mitigation measures on the extant biota. The second route examined the deviation of the biological condition of the HMWB from that of a comparable water body.

The final guidance document on HMWB and AWB recommends the establishment of MEP through the following series of sub-steps (WFD CIS Guidance Document No. 4 2003):

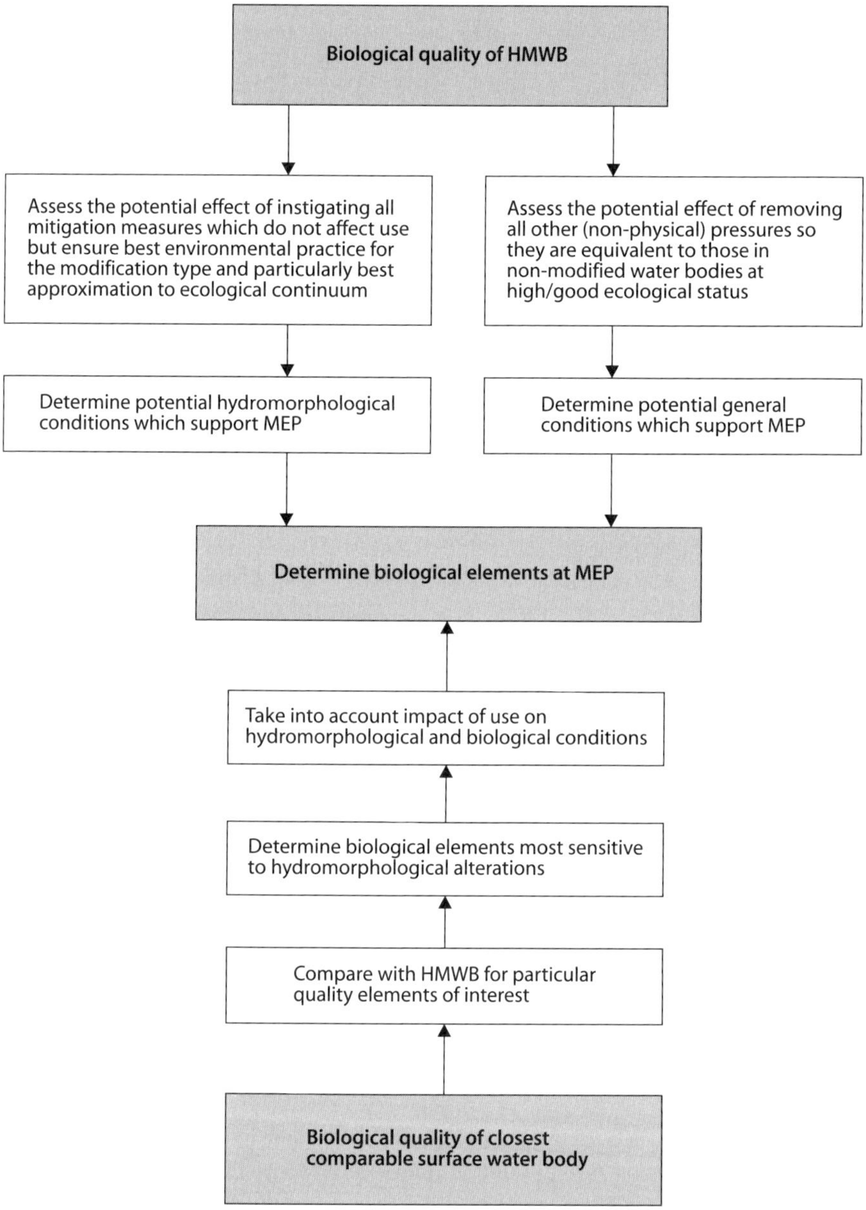

Fig. 9.1. Approach of the UK-S case studies for deriving MEP. *Source:* Case study on the Forth Estuary, UK

- Sub-step 1: Choose the appropriate quality elements for MEP. Identify the closest comparable natural surface water category. This will either be a "river", "lake", "transitional water", or "coastal water". The appropriate quality elements are those of the closest comparable natural surface water category and are identified in WFD Annex V 1.1.1–1.1.4.

- Sub-step 2: Establish the hydromorphological conditions required for MEP. The values for the biological and general physicochemical quality elements at MEP depend on the MEP hydromorphological conditions. Establishing the MEP hydromorphological conditions is one of the first steps in defining MEP since it is these conditions that are impacted by the physical alterations and will, primarily, dictate the ecological potential of an HMWB or AWB.
- Sub-step 3: Establish the MEP physicochemical conditions. Identify the closest comparable surface water body type. Physicochemical conditions at MEP should be based on the conditions of this comparable type-taking account of the MEP hydromorphological conditions. The physicochemical conditions will be an important influence on the values for the biological quality elements at MEP.
- Sub-step 4: Establish the MEP biological conditions that shall reflect, as far as possible, those associated with the closest comparable water body type (cf. sub-step 3 above). The biological conditions at MEP will be influenced by the MEP hydromorphological and physicochemical conditions.

According to the final guidance, in this process it is important to differentiate between "closest comparable surface water category" and "closest comparable surface water body type". The appropriate quality elements are chosen from the closest comparable categories, whereas closest comparable water body types are used to help determine the value of these elements for HMWB and AWB (WFD CIS Guidance Document No. 4 2003).

9.2.1 Mitigation Measures for Achieving MEP

In establishing MEP, Member States are required to define the hydromorphological conditions that would be present if "all mitigation measures" were taken to "ensure the best approximation to ecological continuum, in particular with respect to migration of fauna and appropriate spawning and breeding grounds" (WFD Annex V 1.2.5).

According to the final guidance document on HMWB and AWB (WFD CIS Guidance Document No. 4 2003), mitigation measures for defining MEP should meet the following two main requirements:

a They should not have a significant adverse effect on the specified use or the wider environment; and
b They should ensure the best approximation to ecological continuum, in particular with respect to migration of fauna and approximate spawning and breeding grounds.

The need to meet these two requirements was largely confirmed by the HMWB case studies. In the case studies on the Dender River (B) and Haringvliet Estuary (NL), it was made explicit that measures with adverse effects on the uses or technically infeasible measures were not considered. The authors of the case study on the Rhone River (F) suggested that mitigation measures should be tested also for statutory regulations and financial incentives. In the case study on the Haringvliet Estuary (NL), it remained an open issue whether a cost analysis of the measures for MEP should be carried out in the context of the MEP definition or in the RBMP.

The final guidance document on the HMWB and AWB concluded that the technical feasibility and the financial costs (i.e. capital costs) that would be incurred if the mitigation measures were implemented are not considered in setting the standards for the hydromorphological quality elements at MEP (WFD CIS Guidance Document No. 4 2003). Such cost considerations are relevant when deciding whether the achievement of GEP or a less stringent objective under Article 4(5) of th WFD is appropriate for the HMWB or AWB. However, the mitigation measures should not have a significant adverse effect on the specified use (including economic effects) or the wider environment according to the designation test 4(3)(a). This can include an assessment of the economic effects on the specified use or the wider environment. Although all mitigation measures should be identified, it would not be useful to further consider measures that were impractical. Such impractical measures should be excluded from any detailed assessment (WFD CIS Guidance Document No. 4 2003).

In the HMWB case studies, the definition of the appropriate level of mitigation was based on literature and expert judgment to a large extent. Measures for achieving MEP differed widely depending on the physical alteration, the resulting impact and general condition of the water body (see Table 14.27 in Annex II). Three broad categories of mitigation measures were identified:

- Measures to improve hydromorphological quality (e.g. minimum water level fluctuation, compensation flow, restoration of morphological structures) including measures to ensure the best approximation to ecological continuum (fish passages and creation/restoration of transition zones such as wetlands and littoral zones);
- Measures to improve the physicochemical quality (especially important for lakes/reservoirs): e.g. reduction of diffuse and point discharges, nutrient loads;
- Measures to improve the biological status: habitat improvements, creation of refuges, fish stocking, removal of unwanted species (plants, fish).

The case study on the Dalälven River (S) focused on the sustainability of fish populations in order to define mitigation measures and thereby MEP. Theoretically, reaching MEP would require building well-functioning fish passages where natural fish passage occurred earlier and reducing the storage of water for hydropower purposes. These arguments were presented in a table of criteria (hydromorphological, physical stress factors) together with appropriate thresholds of mitigation for both MEP and GEP (see Table 14.25 in Annex II).

Some case studies were guided by the best environmental practice for the definition of mitigation measures. In the case study on the Danube (A), the definition of mitigation measures was based on state-of-the-art concepts. Similarly, in the case study on the Dee (UK-S), measures were proposed based on elements of good practice established by the hydropower operators. It was, however, pointed out that a code of best practice was needed on appropriate measures to counteract sediment starvation downstream of dams.

Although a selection of the best practice for different types of modifications would be desirable, such a compilation could not be included in this book due to the restricted number of case studies. However, through discussions of the HMWB Working Group, lists of possible mitigation measures could be produced for the specified uses of hydropower, navigation and flood protection (see Tables 9.1–9.3). These can serve as initial checklists that can be consulted with regard to specific uses and physical alterations.

Table 9.1. List of mitigation measures for "hydropower". *Source:* Material presented at the HMWB workshop on 30–31 May 2002 in Berlin

Physical alterations	Mitigation measures
Water flow and water-level fluctuations	– Reintroduce a spring flood – Decrease flow during summer to winter – Avoid flash floods (downstream dams) – Introduce minimum flows (downstream dams)
Habitat improvements	– Terrace tributary outlets (reservoirs) – Reduce erosion in riparian zone – Return boulders to channel (former rapids) – Seed and plant species
Organism and nutrient transport	– Build in fish passages (dams) – Co-ordinated spillway releases (multiple dams) – Remove dam (dams) – Reduce nutrient content in effluents

Table 9.2. List of mitigation measures for "navigation". *Source:* Material presented at the HMWB workshop on 30–31 May 2002 in Berlin

Physical alterations	Mitigation measures
Dams and weirs	– Build fish passages, improve existing fish passages
Channel maintenance/dredging	– Reduce intensity of dredging
Channelisation/straightening	– Connecting existing meanders to the mean channel – Initiating meanders
Bank reinforcement	– More natural embankments
Detachment of oxbow lakes and wetlands	– Cross linking of the river by deepening the oxbow lake – Establishment of natural floodplains
Estuaries	– Restore estuarine conditions (salinity gradient, tidal fluctuation, estuarine morphological processes, migration of fauna)

Table 9.3. List of mitigation measures for "flood protection". *Source:* Compiled by the authors from the evaluation of the HMWB case studies

Physical alterations	Mitigation measures
River straightening, channelisation	– Stream restoration (re-meandering) – Restoration of the catchment
Controlled water level, bank fixation	– Removal of bank fixations – Creation of natural banks and marshes

Only a few case studies such as the case study on Lake Kemijärvi (SF) and the Emån River (S) attempted to determine the exact necessary mitigation measures for achieving MEP (e.g. maximum possible water level fluctuation, minimum required compensation flow, amount of nutrient load reduction).

Nevertheless, several case studies mentioned the expected ecological effects of the proposed mitigation measures (e.g. Dender (B), Loosdrecht (NL), Veluwerandmeren (NL), Kemijärvi (SF), Emån River (S), Great Ouse (UK-EW), and Tummel (UK-S)). Usually, potential effects on the quality elements were assessed based on expert judgment and modelling (e.g. Haringvliet Estuary (NL), Kemijärvi (SF)). In the case of the Dender (B), the ecological effects were roughly estimated by recalculating the modification indicator values and ecological indicators after having applied (in theory) the mitigation measures; this showed that the measures would result in a decrease in the stress factors of erosion, canalisation, banks and locks. It was, however, emphasised that the exact effects of the proposed measures were unknown, similar to other cases such as Haringvliet Estuary (NL), Emån (S), Umealven (S), Tummel (UK-S) and Dee (UK-S). The authors of several case studies mentioned that further research is needed to establish quantitative responses to specific measures. The lack of knowledge and data made it in some cases infeasible to define MEP. In the case of Authie (F), there was a need for further data on the required measures (removal of dams, opening of floodgates and fish ladder installations) and their effects to be able to define MEP.

Ideally, monitoring knowledge should assist in predicting the effects of proposed measures. This was only possible in a few cases such as for the Wienfluss (A) where measures were already in place and gave practical information on their effectiveness.

The HMWB case studies identified the need for interpretation of the term "ecological continuum" used in the WFD (WFD Annex V 1.2.5). It was uncertain how absolute the requirements for ecological continuum are or whether lateral connectivity is part of the ecological continuum.

For the purpose of the final HMWB and AWB guidance document, the "*best approximation to ecological continuum, in particular with respect to migration of fauna and appropriate spawning and breeding grounds*" was interpreted as having the following requirements (WFD CIS Guidance Document No. 4 2003):

- An adequate quantity and quality of usable habitat to ensure that the structure and function of the ecosystem is maintained over space and time.
- Longitudinal and lateral continuity/connectivity of water bodies (e.g. river continuity, aquatic – semi-aquatic – terrestrial habitat connectivity) to enable biota access to the habitats on which they depend.

According to the final guidance document, the best approximation to ecological continuum requires consideration of all hydromorphological mitigation measures that could reduce any obstacles to migration and improve the quality, quantity and range of habitats affected by the physical alterations. This could include connectivity to groundwater and to riparian, shore and intertidal zones. However, the WFD emphasises migration in particular. Priority should therefore be given to reducing any obstacles that significantly inhibit longitudinal and lateral migration of biota.

Regarding mitigation measures in general, the final guidance document on HMWB and AWB concluded that these should:

a Not have a significant adverse effect on the specified use (including maintenance and operation of the specified use). This consideration includes an assessment of possible economic effects incurred by mitigation measures but not an assessment of disproportionate cost of the measures themselves or on the wider environment; and

b Ensure the best approximation to ecological continuum, in particular with respect to migration of fauna and appropriate spawning and breeding grounds.

9.2.2
Quality Elements and Their Values at MEP

Hydromorphological Quality Elements

Hydromorphological quality elements for the definition of MEP were considered in the HMWB case studies as far as they were directly linked to the physical modification or to the mitigation measures. Hydromorphological quality elements included among others:

- Water levels and their fluctuation in reservoirs (Lozoya River (E), Kemijärvi (SF));
- Flow dynamics in regulated rivers (e.g. Suldalslagen (NO), Beiarn (NO, Dalälven (S));
- River continuum (Dalälven (S), Dee (UK-S));
- Degree of fragmentation (Dalälven (S)).

In the case study on the Haringvliet Estuary (NL), the values for hydromorpho-logical quality elements at MEP such as tidal range and maximum current velocity were estimated by ecological modelling. In the Bregenzerach (A), threshold values for maximum daily flow amplitude and maximum daily flow alteration were also estimated.

Chemical and Physicochemical Quality Elements

Physicochemical quality elements seemed to be more important in lakes or reservoirs (Veluwerandmeren (NL), Loosdrecht (NL), Lozoya (E)). In these cases, temperature, oxygen, nutrients and transparency were considered in order to establish MEP. Specific non-synthetic and synthetic pollutants were considered only in a few cases such as the Dee River (UK) and the Haringvliet Estuary (NL) for determining MEP.

Values for the main physicochemical quality elements at MEP were defined only in a few case studies (Lozoya (E), Haringvliet Estuary (NL), Hagmolenbeek (NL)). In the case of the Haringvliet Estuary (NL), no particular reference water body was selected to provide quantitative values for physicochemical elements. Instead, the generic target values used in the Netherlands were applied. Similarly, in Hagmolen-Hegebeek (NL), MEP values for nutrients and oxygen were based on regional standards of negligible risk.

In the case study on the Tummel River (UK-S), the impact of weirs and dams on the temperature regime was investigated. Though the WFD states that "*physicochemical conditions should correspond totally or nearly totally to the undisturbed conditions of the corresponding water body category*", there was evidence for impacts of modifications on the parameters temperature and oxygen.

According to the case study on the Beiarn River (NO), water temperature is altered in many cases due to hydropower schemes. In this case, it was argued that such impacts due to hydromorphological alterations have to be taken into account for the definition of MEP and the related mitigation measures. Chemical problems, however, as in the case of acidification in the Dee (UK-S), should not be included in the definition of MEP. The potential general conditions at MEP should incorporate a theoretical resolution for chemical problems.

The final guidance document on HMWB and AWB recognised that for some AWB and HMWB, the values for some of the physicochemical quality elements in the closest comparable water body type may be significantly different from the values that could be achieved in the HMWB or AWB, given the MEP hydromorphogical characteristics. For instance, the hydromorphological characteristics of a freshwater impoundment created from a dammed estuary may result in different levels of turbidity. These may be different from those in a natural water body. These differences can be taken into account when defining MEP (WFD CIS Guidance Document No. 4 2003).

Moreover, according to the final guidance document, since the values for these physicochemical quality elements would not correspond "totally or even nearly totally to those for the closest comparable water body type" at high ecological status (HES), such AWB and HMWB would never achieve MEP. In some cases, they would also be unable to achieve GEP and therefore would require derogation to a less stringent objective under WFD Article 4(5). Where these physicochemical conditions are directly connected to physical alterations necessary to sustain the specified use, it is suggested that these differences be taken into account when setting MEP. These considerations are only applicable to certain physicochemical elements such as oxygen, temperature and turbidity and should not be applied to general pollutants that are not connected to hydromophological alterations (WFD CIS Guidance Document No. 4 2003).

Biological Quality Elements

Biological quality elements for defining MEP included phytoplankton, aquatic flora, benthic macroinvertebrates and fish fauna. In the case study on the Rhone River (F), only biological quality parameters most sensitive to hydromorphological alterations were used to establish MEP. For the transitional waters of Haringvliet (NL), the area and extent of different estuarine ecotopes (e.g. brackish tidal flats), the salinity gradient as well as the shape of the shores (including modifications) were also used. In the latter case, the values for the quality element of "ecotopes" at MEP were assessed using ecological modelling.

Biological quality elements for defining MEP were usually considered as far as data were available. Difficulties occurred especially with the use of fish fauna for defining MEP. In the case study on the Lozoya River (E), this was due to the general lack of natural lakes that could serve as appropriate reference sites for the reservoirs in question. In the case study on the Forth Estuary (UK-S), fish fauna was excluded from the definition of MEP due to the high annual variation of the fish population. In the case study on Lake Kemijärvi (SF), it was noted that fishing and fish stocking were not taken into account in the WFD text. Therefore, if the status of the fish fauna is deteriorated due to the intensive fishing of predatory fish, the ecological status should not be lowered. Thus, the Finnish case study pointed out that it is essential to distinguish between changes caused by water quality problems or hydro-morphological alterations from changes caused by fishing or fish stocking.

The values for biological quality elements at MEP were only determined in a few case studies. An example was the case study on Lozoya (E), where the maximum density of phytoplankton was determined. In Box 9.2 and Table 9.4, the option of hindcasting modelling to set MEP values in Lake Loosdrecht (NL) is presented. Usually only the general trend of species diversity and abundance (increase/decrease) or community structure (e.g. dominant macrophytes) at MEP was predicted. The methods used to achieve this varied. In the case of Haringvliet (NL), information on species at MEP was based on expert judgment and historical data. In the Dender (B), conclusions on the presence of species were based on the ecological indicator list of the stream typology system used in the area.

In some cases, it was assumed that if the hydromorphological and physicochemical elements meet the values at MEP, the associated biological status represents the biological quality elements at MEP. Therefore, in some case studies the biological quality elements at MEP themselves were not defined.

In general, case studies using the second approach for defining MEP with an emphasis on mitigation measures (see Sect. 9.2 of this chapter) did not succeed in defining values for the quality elements. This was mainly due to the fact that the ecological effects of each mitigation measure and their combination were difficult to predict. Due to little information about the interaction between water quality, habitat quality and the presence of species, it was difficult to predict the response of biology to the mitigation measures necessary. In several case studies, the need was expressed for further research on the effects of measures in order to define MEP appropriately.

Box 9.2. Hindcasting Modelling to Set MEP Values in Lake Loosdrecht, The Netherlands

Lake Loosdrecht (10 km^2) was selected as a typical example for a Dutch peat lake and is part of a system of shallow interconnected lakes in the centre of the Netherlands. While former industrial peat mining created the preconditions for the creation of the lake, natural processes formed the present lake. It is heavily impacted by recreation, especially recreational shipping and by the construction of unnatural embankments to prevent further erosion of the peat banks. A further problem is eutrophication and water quality.

The HMWB designation showed that a return to the oligotrophic reference phase is not possible for Lake Loosdrecht. Therefore, the Maximum Ecological Potential was based on Phase 2 in the history of Lake Loosdrecht: the clear, mesotrophic lake dominated by macrophytes. The MEP was described on the basis of historical data from Phase 2 (1930–1955) in the history of Lake Loosdrecht. This phase is the period after the oligotrophic phase, which is the reference condition of the lake, and it is the phase before the eutrophication of the lake into a turbid, cyanobacteria-dominated situation. The values of the biological and physicochemical parameters during Phase 2 were derived from the AMOEBE Loosdrecht. The AMOEBE is an ecological assessment instrument, which describes the ecological objective of Lake Loosdrecht (see Table 9.4). The desired values presented in Table 9.4 corresponded to the values required for MEP. They were not based on measurements but were estimated on the basis of historical information on biological elements. Therefore the approach used in this HMWB case study for the establishment of MEP biological requirements involved a hindcasting method using historical data.

Source: Case study on Lake Loosdrecht, NL.

Table 9.4. Values at MEP of the biological and physicochemical parameters of Lake Loosdrecht, The Netherlands. *Source:* Case study on Lake Loosdrecht, NL

Parameter	Present value (1990)	Desired value
Abiotic		
Transparency (yearly average, m)	0.4	1.9
Total P (yearly average, mg P l^{-1})	0.1	0.0054
Soluble reactive P (yearly average, mg P l^{-1})	0.002	0.0015
Mineral nitrogen (summer average, mg N l^{-1})	0.840	0.238
Salinity (summer average, mg l^{-1})	305	231
Oxygen (summer average, mg O_2 l^{-1})	10.4	11.0
Plankton		
Total zooplankton (carbon, mg C l^{-1})	0.575	0.075
Total cyanobacteria (fresh weight, mg l^{-1})	30.6	1.25
Total diatoms (fresh weight, mg l^{-1})	0.81	0.06
Total green algae (fresh weight, mg l^{-1})	0	0.48
Chlorophyll a (mg l^{-1})	0.1	0.025
Vegetation		*total covering according to Tansley (1946)*
Characeae-group (4 species)	0	15
Potamegoton-group (10 species)	1	11
Menyanthes-group (7 species)	2	14
Thelypteris-group (15 species)	6	25
Nymphaea-group (7 species)	14	18
Butomus-group (6 species)	11	15
Caltha-group (10 species)	14	24
Fish		
Pike (fresh weight, kg ha^{-1})	1	45
Bream (fresh weight, kg ha^{-1})	180	50
Surroundings of lake		
Presence of helophytes and marsh vegetation	Absent due to fixed banks	Presence due to the existence of natural banks and marshes

In the UK-EW case studies, the RIVPACS system was recommended for generating predicted macroinvertebrate assemblages for streams and rivers. However, according to the authors of these case studies, in its present form RIVPACS is designed to respond most strongly to water quality variables, while the influence of these variables are inseparable from the effects of physical modification. Assessment of macro-invertebrate responses to physical modification requires a suitable range of reference sites with matching physicochemical characteristics and good water quality status (to maximise the macroinvertebrate diversity and hence the “signal”).

In the case of the Dee (UK-S), a suitable spatial network of reference sites was chosen for targeted sampling of macroinvertebrates at sites impacted by hydropower as well as unimpacted reference sites. This strategy was followed because of the lack of biological standards and the unavailability of the Scottish reference typology at the time.

In the case study on the Tame River (UK-EW), it was suggested that the values of fish fauna at MEP should be defined as fish assemblages (densities, biomass and species diversity) similar to those of water bodies with similar modifications and with the "best management practice".

According to the final guidance document on HMWB and AWB, the WFD allows a number of methods to be used in establishing MEP values for the biological quality elements. The methods are the same as those permitted in establishing the values for quality elements at HES. They consist of spatial networks of sites meeting MEP criteria, modelling approaches or a combination of those. Where it is not possible to use the above methods, expert judgment may be used (WFD CIS Guidance Document No. 4 2003).

Identification of a Closest Comparable Surface Water Body

As part of the establishment of MEP, the identification of a "comparable water body" helps to:

- Choose quality elements to be regarded (derived from the most comparable water body *category*); and
- Set values for physicochemical and biological quality elements regarded (derived from the most comparable water body *type*) (WFD CIS Guidance Document No. 4 2003).

A "comparable water body" can be one or more similar water body(s) that is/are, amongst other things, most similar in terms of category, type and other characteristics to the modified water body and from which spatial or temporal (i.e. hindcasting) data can be derived to support the establishment of MEP (WFD CIS Guidance Document No. 4 2003).

The final guidance document on HMWB and AWB sets as the first priority to look for a comparable *natural* water body (or a modelled or historical situation). With respect to this, the approaches followed in the case studies can be summarised as follows (see Table 14.28 in Annex II):

i The majority of the HMWB case studies used *natural (unmodified) water bodies* as comparable water bodies. For instance, pristine reference streams were considered in the case study on the Umealven River (S), where the free-flowing Vindel River running parallel to Umealven served as reference. In the case study on the Tummel River (UK-S), single pristine stretches were selected. In the Dee (UK), natural analogues were used for water bodies both with and without a change in water category; for the latter, the DHRAM method (Dundee Hydrologic Regime Alteration Method)[1]

[1] DHRAM has been developed as a tool for assessing the severity of hydrologic regime alteration on a five-class scale, using the concept of risk to indicate the likely severity of regime alteration.

was used with a set of natural Scottish lochs as reference. The German approach was also to use natural waters as reference arguing that in the case of comparison with other modified water bodies, it would be necessary to define MEP for every water category and type depending on the use.

ii Few case studies used *modified water bodies* (with respect to the specific physical alterations) as comparable water bodies. The Swedish case study on the Emån River as well as the UK-EW case studies used "similar physical alterations and anthropogenic pressures" as a criterion for identifying comparable water bodies.

iii Several case studies used both *modified and natural water bodies* as comparable. The Lozoya River case study (E), for instance, considered only one natural lake as reference; the remaining comparable water bodies were four reservoirs of the same typology due to the lack of natural lakes. In the case of the Rhone River (F), comparison to both natural and modified water bodies was attempted, but MEP was finally determined through reference to modified water bodies. The case studies on the Haringvliet Estuary and the Hagmolenbeek River (NL) also made use of both natural and modified water bodies in the search for comparable water bodies. For Haringvliet (NL), several national and European modified and unmodified estuaries were compared with regard to their total volume, loss of intertidal area and estuary volume index. However, no comparable estuary to the Haringvliet could be identified.

Through the evaluation of the HMWB case studies, it became obvious that comparable water bodies in practice were hard to find. Seven case studies (Bregenzerach and Danube (A), Dender (B), Dalälven and Baltic coast (S), Veluwerandmeren (NL), and Beiarn (NO)) were not successful in finding comparable water bodies. In the case of the Baltic coast (S), comparable water bodies could not be found due to the high natural variability of the ecosystem in question and the lack of unaffected areas.

In the case of the Forth Estuary (UK-S), no comparable body could be found due to natural variability, but an analogue could be derived from a model based on the estuary itself; the boundaries of the estuary were modified to represent the condition at MEP. This method helped to avoid typological discrepancies that would be introduced by using another estuary. Similarly, the case studies on Lake Kemijärvi (SF) and the Wienfluss (A) used the historical natural situation of the water bodies themselves as reference instead of searching for other comparable water bodies.

In this context, the case study on the Hagmolenbeek River (NL) argued that it is a better option to derive MEP from the reconstruction of a theoretical status using ecological prediction models, historical data or spatial reference waters. It was considered risky to base the definition of MEP on the characteristics of an existing comparable water body, since two streams can only be comparable to a certain extent.

As mentioned above, spatial networks, modelling and historical situations were among the options used to find comparable water bodies. Other criteria used for the selection of comparable water bodies included the presence of:

- Similar anthropogenic pressures and uses;
- Corresponding physical and hydromorphological characteristics such as altitude, geology, length/size, catchment area, depth and geographic proximity, typology, and stream order;

- Comparable biological data. With regard to this, the case study on the Lagan River (UK-NI) also mentioned that for comparative purposes, the "best of the type" of similar rivers in terms of quality of biological elements should be chosen.

The final guidance document on HMWB and AWB recognised that in many cases, the High Ecological Status (HES) hydromorphological and sometimes also the physicochemical conditions in the closest comparable water body type will be significantly different from the MEP hydromorphological and physicochemical conditions (WFD CIS Guidance Document No. 4 2003). In establishing the MEP biological values, it will therefore be necessary to adjust the HES biological values of the closest comparable water body type to take account of the heavily modified or artificial characteristics. In special cases, comparable natural water bodies will not be available. In these cases, which have to be justified, information from closely comparable HMWB and AWB at MEP (i.e. the best possible rather than best available) should be used where it is available. Information from the best available sites could be used as long as the best possible conditions can be extrapolated through modelling or expert judgment (WFD CIS Guidance Document No. 4 2003).

9.3 Conclusions and Discussion

The main conclusions and questions concerning the definition of MEP are the following:

- Two different approaches for defining MEP were identified in the HMWB case studies: According to the first approach, the definition of MEP was based on reference conditions similarly to the definition of "high ecological status" for natural waters, while the second approach emphasised the mitigation measures. In some case studies, a combination of both approaches was used for the definition of MEP. The first approach was followed by the majority of the HMWB case studies and is in line with the final guidance document on HMWB and AWB. The final guidance document also emphasised the importance of mitigation measures, which have been the focus of the second approach as part of the definition of the hydromorphological conditions at MEP. Establishing these hydromorphological conditions is one of the first steps in defining MEP since it is these conditions that are impacted by the physical alterations and will, primarily, dictate the ecological potential of an HMWB or AWB.
- Regarding the identification of mitigation measures for achieving MEP, the need was expressed for "best environmental practice" guidance for different types of modification related to different uses. Within the scope of this book, initial check lists of mitigation measures could be produced for the uses of hydropower, navigation and flood protection (see Sect. 9.2.1 of this chapter).
- Several case studies assessed the expected ecological effects of the proposed mitigation measures. Usually, potential effects on the quality elements were assessed based on expert judgment and modelling. It was however often difficult to predict the (quantitative) response of biology to specific mitigation measures necessary for achieving MEP (and later GEP). Several case study authors expressed the need for further research on the effects of mitigation measures in order to appropriately define MEP.

- Although the WFD states that (at MEP) "*physicochemical conditions should correspond totally or nearly totally to the undisturbed conditions of the corresponding water body category*", there was evidence for impacts of hydromorphological alterations on the parameters temperature and oxygen (e.g. in the case of hydropower production, water temperature is affected). Some HMWB case studies suggested taking such impact on the physicochemical parameters into account for the definition of MEP.
- Through the evaluation of the HMWB case studies, it has become obvious that comparable water bodies in practice are hard to find. Most case studies used *natural (unmodified) water bodies* and few used *modified ones* as comparable water bodies. Use of modified water bodies was made mainly in case of lack of natural ones corresponding to the characteristics of the water body in question.

Chapter 10

Definition of Good Ecological Potential

10.1 Introduction

As shown in the Heavily Modified and Artificial Water Bodies (HMWB and AWB) designation process (Fig. 2.1, p. 12), the establishment of Maximum Ecological Potential (MEP) is followed by the establishment of Good Ecological Potential (GEP) (Step 11 of Fig. 2.1). Good Ecological Potential (GEP) and good chemical status are the environmental objectives for AWB and HMWB, and they are to be achieved at the latest by December 2015 (Art. 4(1)(a)(iii)). GEP takes account of the constraints imposed by the effect of physical alterations upon a water body's hydromorphology and consequently upon its biology. GEP is equivalent to achieving Good Ecological Status in unmodified water bodies once the impact of physical modifications necessary to support the use has been taken into account.

According to the WFD (WFD Annex V 1.2.5), Good Ecological Potential is defined as the state where "*there are slight changes in the values of the relevant biological quality elements as compared to the values found at Maximum Ecological Potential*". The hydromorphological conditions at GEP are expected to be consistent with the achievement of the values specified for the biological quality elements. Physicochemical and nutrient conditions are within the ranges established to ensure the functioning of the ecosystem and the achievement of the values specified for the biological quality elements.

In this chapter, the results and approaches of the HMWB case studies relevant to the definition of GEP are presented.

10.2 Methods for the Definition of GEP

In summary, two different approaches were identified for the definition of GEP in the HMWB case studies (see Table 14.29 in Annex II):

a GEP was defined by defining the "slight changes" of GEP from MEP with regard to the values of the biological quality elements (Bregenzerach (A), Ruhr (D), Lozoya (E), Authie (F), Dalälven (S), Tame (UK-EW), Forth Estuary (UK-S), Dee (UK-S), and Tummel (UK-S)).

b GEP was defined by selecting the mitigation measures necessary for achieving MEP and taking into account their practicability and realisation (Danube (A), Dender (B), Sarre (F), Suldalslagen (NO), Beiarn (NO), Umealven (S), and Emån (S)). This approach included consideration of cost-effectiveness as well as technical feasibility of measures.

Approach (a) involved interesting discussions on the interpretation of the term "slight changes":

- The case study on the Lozoya River (E) defined "slight changes" for each biological quality element and also defined the relevant hydromorphological and physico-chemical conditions at GEP. In this case, GEP and the values of the relevant quality elements corresponded to a mesotrophic condition of the reservoirs as compared to the oligotrophic condition at MEP.
- Quite often expert judgment was used to define the term "slight changes" but there were also cases where more specific parameters were applied. For instance, in the case of the Authie (F), the standardised global biological index of a value 18/20 was used to define the state of GEP (20/20 being equivalent to MEP). In the UK-EW guidelines (Booker et al. 2002a), the level of deviation of GEP from MEP was set at 2–25%; this was based on deviation from reference assemblages using the following quality elements: habitats, fish, plants, and benthic invertebrate fauna.
- According to the case studies on Lake Loosdrecht and Lake Veluwerandmeren (NL), the relationship between abiotic parameters and ecological functioning of shallow lakes is not linear but shows two equilibrium states (clear and turbid), which can switch depending on the nutrient loading, restoration measures or the area of macrophytes and helophytes. In the case of Lake Veluwerandmeren, the MEP was equal to an equilibrium state of a clear lake. However, abiotic values, which were somewhat higher or lower than the ones at MEP, could lead to a turbid state of the lake, in which the biological parameters deviated largely from the ones at MEP. Therefore, for both cases on the Dutch lakes, it was decided to set the same quality element values for GEP as for MEP. The difference between MEP and GEP was in the different ecological condition of the banks. In general, at MEP the banks were natural, whereas at GEP they remain fixed.

Regarding *approach (b)*, the following can be summarised:

- The definition of GEP in the case study on the Haringvliet Estuary (NL) was based on the practicability of the measures identified for achieving MEP. The cost effectiveness and restoration effect of different scenarios were compared. The relationship between GEP/MEP/GES/HES (WFD objectives), costs and the percentage of ecological restoration can be seen in Fig. 10.1. High ecological status (HES) was considered 100% ecological recovery and the present situation no recovery. Figure 10.1 illustrates that for GEP the ecological restoration achieved is quite high already (70–80%), while the costs are almost 50% lower than those to be invested for MEP. In this scenario, GEP can, therefore, be considered a slight ecological change to MEP (80% ecological restoration instead of 90%), in which the ecological continuum especially for fish migration is restored (sluices are still opened) and in which costs are much lower than for MEP.
- Similaly, in the case study on the Sarre River (F), the parameter values at GEP should reach a minimum level of 60% of HES after mitigation. This could be reduced to 50% when many disturbances accumulate in one river section.
- In the case of the Danube (A), the technical feasibility and costs of the mitigation measures were defined by a number of implemented or intended restoration programmes.

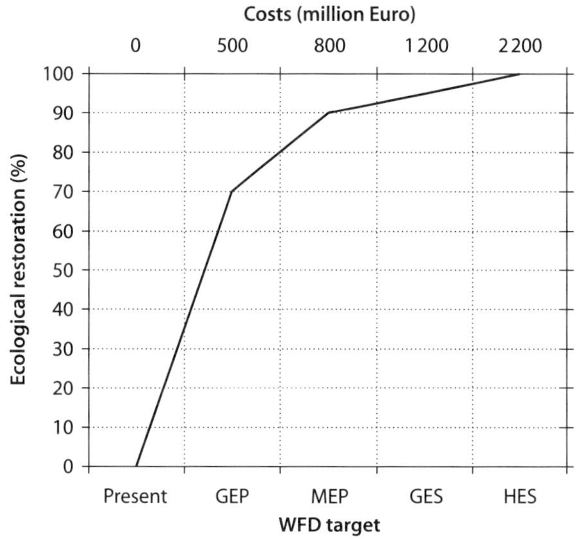

Fig. 10.1. Relationship between WFD objectives, costs and percentage of ecological restoration, The Netherlands. *Source:* Case study on the Haringvliet Estuary, NL

- In the case studies on the rivers Suldalslagen (NO), Beiarn (NO), Emån (S) and Umealven (S), GEP was determined as the status that is achieved when all mitigation measures for achieving MEP are realised, except for those which are assumed to be impracticable or disproportionately costly. In the case studies on the rivers Suldalslagen and Emån, the construction of fish passages was considered to be too expensive and was therefore not considered for the definition of GEP. In the case study on the Umealven River (S), all mitigation measures for achieving MEP were considered for the definition of GEP, except for those that would lead to >10% loss of hydropower production. Defining values for the quality elements at GEP was not attempted in any of these case studies.
- The case study on the Kennet (UK-EW) pointed out that the difference between MEP and GEP was somewhat arbitrary, and therefore a pragmatic approach was adopted. The authors of this case study interpreted GEP as adopting the best practice in river management given the modifications necessary for the uses. They further argued that this interpretation can give clear guidance to ecologists on what mitigation measures to suggest.
- In the case study on the Forth Estuary (UK-S), although approach *(a)* as mentioned in the beginning of this section was followed, it was noted that there are uncertainties in the biological interpretations of MEP and therefore GEP. The authors stated that an alternative approach based on economic analysis may be worth exploring (similar to approach *(b)*). This approach would define practicable mitigation measures in terms of a cost threshold that would then be related to a minimum level of biological quality for the status of GEP.

The final guidance document on HMWB and AWB (Sect. 6.3 of the WFD CIS Guidance Document No. 4 2003) recommended the establishment of GEP through the following series of sub-steps:

- Sub-step 1: The establishment of Good Ecological Potential for HMWB and AWB is principally based on the *biological quality* elements (derived from MEP). GEP accommodates "slight changes" in the values of the biological elements from the MEP.
- Sub-step 2: The *hydromorphological conditions* at GEP must be such as to support the achievement of the GEP biological values. This will require the identification of the hydromorphological conditions necessary to support the achievement of the GEP values for the biological quality elements, and in particular the achievement of the values for those biological quality elements that are sensitive to hydromorphological alterations.
- Sub-step 3: The values for the *general physicochemical* quality elements at GEP are such that they support the achievement of the GEP biological values. It is also required that the values for the general physicochemical quality elements at GEP are such that they ensure the functioning of the ecosystem (WFD Annex V 1.2.5).
- Sub-step 4: GEP also requires compliance with environmental quality standards established for the *specific synthetic and non-synthetic pollutant* quality elements in accordance with the procedure set out in WFD Annex V (1.2.6).

Approach (b) of the HMWB case studies as mentioned in the beginning of this section is not in agreement with the final guidance document on HMWB and AWB but seems more relevant to the establishment of the Programme of Measures (POM) for protecting and enhancing ecological quality (see Sect. 10.4 of this book).

Calculations of Ecological Quality Ratios

The Water Framework Directive requires the reporting of Ecological Quality Ratios (*EQR*), which represent the relationship between the values of the biological parameters observed for a given body of surface water and the values for these parameters in the reference condition applicable to that body (WFD Annex V 1.4). The ratio is expressed as a numerical value between zero and one, with higher ecological status or potential represented by the higher number.

EQR were only calculated in very few of the HMWB case studies. The approaches used are presented here.

In the case study on the Dender River (B), *EQR* were calculated as the ratio of the Belgian Biotic Index (*BBI*) values. The formula thereby used was $EQR(\text{BBI})$ = actual *BBI* / needed *BBI*.

The UK-S case studies also calculated Ecological Quality Ratios (*EQR*) in order to derive GEP from MEP. In the case study on the Forth Estuary (UK-S), *EQR* were derived for each biological quality element (and habitat combinations) and the results were combined to derive the *EQR* for the water body as a whole. In the case study on the Tummel River (UK-S), the calculation of *EQR* was based on the available benthic invertebrate fauna and macrophyte data. In order to derive GEP from MEP, the UK-S case studies divided the scale of *EQR* into four equal intervals corresponding to the levels of ecological potential as shown in Table 10.1.

A specific form for the calculation of *EQR* was used: $EQR = C / (M + C + E)$, where C is taxa common with those of the reference condition, E is "exotic" taxa not recorded at reference condition and M are reference taxa which are missing.

Table 10.1. Ecological potential and ecological quality ratios in the UK-S case studies. *Source:* Case study on the Tummel River, UK-S

Ecological Potential	*EQR*
Maximum	1.00
Good	0.75–0.99
Moderate	0.50–0.74
Poor	0.25–0.49
Bad	0.00–0.24

However, the authors of the UK-S case studies noted that the proposed division of the *EQR* scale required verification and calibration with further biological data. The threshold value of 0.75 was a nominal value and was likely to require adjustment in the light of more detailed assessments on the effects of various measures on ecology.

In the case of the Dee River (UK-S), it was sometimes not possible to say whether a low *EQR* was the effect of acidification or sediment starvation. According to the case study authors, this would be possible only when a non-acidified reference is found. In order to deal with such cases, an alternative was proposed to derive the *EQR* as the reciprocal of risk score (and thereby calculate separately *EQR*s relevant to GES and *EQR*s relevant to MEP). *EQR* relevant to MEP (physically-based *EQR*) were obtained by recalculating the risk score to include only factors that could be removed (through "all mitigation measures") without prejudice to the designated use.

10.3 Results

In some case studies, the current status of the HMWB was classified in terms of the current ecological potential (see Table 10.2). The Tame River (UK-EW) regarding fish population and Lake Kemijärvi (SF), the Suldalslagen River (NO) and the Ruhr River (D) were already considered to be at Good Ecological Potential. In the case of Lake Kemijärvi (SF), various measures were already in place. This was also the case in the Wienfluss (A) where, however, measures should be extended to the entire river for GEP to be achieved.

Apart from hydropower regulation, the Suldalslagen River (NO) was not much affected by anthropogenic activities. Moreover, various mitigation measures were already in place, and the Suldalslagen was therefore considered not to deviate much from MEP in terms of the biological quality elements (in spite of the lack of anadromous fish). For the Tummel River (UK-S), the current ecological potential was classified as "moderate to good" and for the Forth Estuary (UK-S) as "moderate". In the Lozoya River (E), a further improvement to the current eutrophic-mesotrophic status would be necessary to reach GEP.

10.4 Programme of Measures for Protecting and Enhancing Ecological Quality

Article 11 of the Water Framework Directive requires the establishment of a Programme of Measures (POM) designed to achieve the objectives of the Directive, i.e. to

Table 10.2. Results on current ecological potential. *Source:* Compiled by the authors from the evaluation of the HMWB case studies

Case study	Current ecological potential
Wienfluss (A)	GEP could be achieved when the rehabilitation measures taken so far are extended to the entire Wienfluss
Ruhr (D)	GEP
River Lozoya (E)	Mesotrophic to eutrophic status < GEP
Lake Kemijärvi (SF)	MEP would not differ significantly from current status, therefore GEP assumed
River Suldalslagen (NO)	Biological quality elements: MEP Hydromorphological quality elements: GEP Physicochemical elements/pollutants: GEP
Tame (UK-E&W)	GEP for fish population
River Tummel (UK-S)	Moderate to good ecological potential
Forth Estuary (UK-S)	Moderate ecological potential

achieve GEP in the case of HMWB and AWB. The Programme of Measures includes basic and supplementary measures.

Basic measures are those that are required to implement EU legislation for the protection of water, including the various directives listed in Part A of Annex VI of the WFD. For HMWB, measures will include those required for the achievement of Good Ecological Potential.

In many of the HMWB case studies, the mitigation measures identified for achieving GEP (see Table 14.29 in Annex II) were also considered as the basic measures of the POM because they aim at enhancing actual ecological quality. The link between GEP and measures themselves was based on the present understanding of physical factors that are known or thought to exercise control on the ecological status.

In the case of the Dender River (B), basic measures were considered to be relevant to hydromorphological factors contrary to supplementary measures that were considered to be directed at water quality. In some cases like that of the Forth Estuary (UK-S), basic measures were related to EU directives including the Waste Water Treatment Directive, the Nutrients Directive, and the Habitats and Birds Directive (see Table 14.30 in Annex II for details on the basic measures of the POM).

In some HMWB case studies, the Programme of Measures was based to a large extent on action plans, water development and management plans (existing or in preparation) regarding the water bodies in question (as in the cases of the Rhone River (F), Veluwerandmeren (NL), and Lozoya River (E)). In general, basic measures included the following:

- Morphological restoration measures (e.g. construction of fish passages);
- Hydrological restoration measures (e.g. abstraction controls, minimum tapping to protect bird species);
- Water quality measures (e.g. controls for point source discharges and diffuse sources including wetland creation and construction of wastewater treatment plants); and
- Other measures (e.g. re-introduction of species, use of requirements of Natura 2000).

It is not considered helpful to summarise all measures proposed in the HMWB case studies for the purposes of this book, since the Programme of Measures is determined on a case-by-case basis. Only the POM proposed in the case study on the Emån River (S) is presented in detail in Box 10.1 as an example.

Supplementary measures such as those listed in Part B of Annex VI of the WFD may also be employed where appropriate in order to protect and enhance ecological potential. These supplementary measures may be important in increasing the probability of achieving environmental objectives in situations where basic measures do not guarantee such outcomes.

In the HMWB case studies, supplementary measures included, among others, measures targeted towards institutional instruments such as measures involving the content of the River Basin Management Plan and negotiated agreements on the management of reservoir operations in the Lozoya River (E) (see Table 14.30 in Annex II for details on the supplementary measures of the POM). Supplementary measures also consisted in promotion of codes of good practice: promotion of better agricultural practices in the case of the Lozoya River (E) and Authie River (F) and collaboration with electricity operators for a code of practice on environmental impact minimisation from abstraction in the case of the Forth Estuary (UK-S). Other issues considered in the definition of supplementary measures were active information of stakeholders (e.g. in the Authie River (F)) as well as the possibility for environmental certificates for hydropower plants linked to the WFD GEP requirements (Emån River (S)).

In some cases, it was not possible to distinguish between basic and supplementary measures (River Ruhr (D), Hagmolenbeek River (NL) and Lake Loosdrecht (NL)). In the case of Hagmolenbeek (NL), it was noted that the difference between the two types of measures remained unclear.

According to the final guidance document on HMWB and AWB, the establishment of the Programme of Measures requires a good understanding of how measures will improve the ecological potential of the water body (WFD CIS Guidance Document No. 4 2003). For example, the identification of the relevant GEP hydromorphological conditions will require an understanding of the relationships between hydromorphological and biological elements; this knowledge is still relatively limited. It would also be advantageous to understand biological response lag times within any particular water body. The final guidance pointed out that for the design of effective and efficient programmes of measures (POMs), better information is likely to be collected over time. In the meantime, Member States will have to base the design of POMs on the best available knowledge and judgments (WFD CIS Guidance Document No. 4 2003).

10.5
Conclusions and Discussion

The main conclusions and questions on the definition of GEP are the following:

- In summary, two different approaches were identified for the definition of GEP. The first one was based on the definition of the term "slight changes" of GEP from MEP with regard to the values of the biological quality elements. The second one was based on a selection of the mitigation measures necessary for achieving MEP taking into account their practicability and realisation. The second approach of the HMWB case

Box 10.1. Measures for Achieving GEP in the Emån River, Sweden

The Emån River is the largest river in south-eastern Sweden and is heavily affected by over forty hydropower plants along its course. In the sub-basin (River Solgenån) where the case study was focused, there were three large hydropower plants, hindering the migration of *Salmo trutta*. Here the proposed set of measures (basic and supplementary) for achieving Good Ecological Potential is outlined.

Basic Measures

According to the WFD, the relevant basic measures for this case study were surveillance of pressures on the hydromorphology, recovery of costs for water services and the measures necessary to fulfil the requirements of the directives listed in WFD Annex VI Part A. Out of the directives mentioned in WFD Annex VI Part A, only those considering physical and hydromorphological modification were regarded in this case study, since only those kinds of modifications were studied. The following measures were considered to fulfil the requirements of the WFD:

- Careful analysis before more water withdrawals are allowed. Both the Water Framework Directive and the Environment Impact Assessment Directive should be used for the assessment of the impacts of new water withdrawals. It is important to find out whether a new water withdrawal would threaten the achievement of Good Ecological Potential. Since the highest possible minimum tapping of 5% is much lower than the ecologically optimal minimum release, there is a risk that a water withdrawal might have a considerable effect in this area.
- The costs for the water services could be gathered in trust funds. The hydropower companies could pay an amount of money each year or as a lump sum. These trust funds could finance for example, ecosystem restoration and protection, and public education. It is suggested that these funds be managed and the money distributed by an advisory council with members such as environmental authorities, NGOs and others concerned.
- According to the Birds Directive (79/409/EEC) Article 4: "*The species mentioned in Annex I shall be the subject of special conservation measures concerning their habitat in order to ensure their survival and reproduction in their area of distribution*". Out of the bird species mentioned in Annex I, the following species are likely to be found in one water body of the River Solgenån: black-throated diver, whooper swan, smew, osprey, crane, wood sandpiper, common tern and kingfisher. The mitigation measures that would be necessary in order to protect these species are a minimum tapping to improve the environment in the old, dry river channels and gradual increases/de-creases when the tapping is changed.
- The aim of the Habitats Directive (92/43/EEC) is "to contribute towards ensuring biodiversity through the conservation of natural habitats and of wild fauna and flora". A European ecological network of areas is, at the moment, set up under the title of Natura 2000. River Solgenån (lower) has not been chosen to be one of the Natura 2000 areas, but Lake Solgen is a Natura 2000 area. According to the Water Framework Directive Article 4.8, the measurements to achieve Good Ecological Potential must not risk jeopardising the achievement of the aims of the WFD or other EU-legislation in adjacent waters. As a consequence, the mitigation measures in water body 3 (of the River Solgenån) to achieve Good Ecological Potential must not put the aims of the Habitat Directive in water body 1 (Lake Solgen) at risk.

Supplementary Measures

The definition of supplementary measures is found in the WFD Article 11.4 and Part B of WFD Annex VI. The following supplementary measures were suggested for this case study:

- The criteria for environmental management standards such as ISO 14 000 could in the case of the certification of hydropower companies be complemented with the requirements of the WFD. It is suggested that MEP should be fulfilled for those hydropower plants situated in areas classified as HMWB in order to achieve an environmental certificate. For those hydropower plants situated in areas not classified as HMWB, at least Good Ecological Status should be achieved and maintained in order to achieve an environmental certificate.

Box 10.1. *Continued*

- A very important part of the work with the WFD is to inform the public. One of the sub-catchments of the Emån River could be used as a demonstration site for public information and involvement. This area could be restored and carefully monitored in order to investigate the effects of the restorations. Such a project could be a co-operation between the county administrative board and district authorities in the study area such as already existing networks like the Emå-project and the Emån water society, NGOs and others. Apart from demonstration of river restoration and investigation of the effects, the aims of such a project should be to inform the public about the WFD, the pressures on rivers and what can be done. The project could be financed by money from the trust funds suggested as one of the basic measures.
- Time-limited licenses for hydropower plants are used in the US. These licences include environmental requirements and have to be renewed every thirty to fifty years. When an application for relicensing is made, the licensee has to provide a description of the financing and the environmental impacts of the project. Both the concerned authorities and the public get an opportunity to review the plan and, for example, suggest additional measurements to mitigate the environmental impacts. It is suggested that this method should be used also in the case of the Emån and in EU countries in general.

Water-saving measures for industries and agriculture could also be useful in order to maintain Good Ecological Potential. The regulation of Lake Solgen does not only function as a reservoir for the hydropower plants but also as one of the reservoirs for the rest of the Emån River. Water-saving measurements could, for that reason, result in a lower regulation of Lake Solgen and more water in Solgenån (lower) and a better environment in water body 3 (of the River Solgenån).

Source: Case study on the Emån River, S.

studies is not in agreement with the final guidance document on HMWB and AWB but seems more relevant to the establishment of the Programme of Measures (POM) for protecting and enhancing ecological quality.

- GEP could only be defined on a weak basis due to uncertainty with regard to the overall HMWB designation process and with regard to the definition of MEP. As a result of the difficulty encountered in defining MEP, it was equally difficult to define GEP and "slight changes" from MEP.
- Ecological Quality Ratios (*EQR*) were only calculated in a few of the HMWB case studies. The need was expressed for appropriate biological monitoring data in order to calculate meaningful *EQR* values.
- The need was identified to clarify how and when to deal with the costs of measures necessary for achieving MEP and GEP. Two possibilities were identified in the HMWB case studies: in the process of defining MEP and GEP or as part of the River Basin Management Plan.

As already mentioned, according to the final HMWB and AWB guidance document, the technical feasibility and the financial costs (i.e. capital costs) that would be incurred if the mitigation measures were implemented is not a consideration in setting the standards for the hydromorphological quality elements at MEP (and GEP) (WFD CIS Guidance Document No. 4 2003). Such cost considerations are relevant when deciding whether the achievement of GEP or a less stringent objective under Article 4(5) is appropriate for the HMWB or AWB. For the Programme of Measures to achieve the environmental objectives, i.e. GEP for HMWB and AWB, costs of measures (including disproportionality of costs) should be considered. Table 10.3 of the final guidance document provides a relevant overview of measures and cost considerations in the overall HMWB and AWB designation process. The

Table 10.3. Overview of measures and cost considerations in the final HMWB and AWB identification and designation process. *Source:* WFD CIS Guidance Document No. 4, 2003

Step	Measures to be considered	Costs (and benefits) related to measures/other means
1–6: Up to provisional identification	None	Not considered
7: Designation test 4(3)(a)	Restoration measures necessary to achieve GES	– When assessing the adverse effects on the specified uses and on the wider environment, costs need to be considered – The benefits of achieving GES must be considered; other benefits may be considered – Costs of restoration measures (including disproportionality of costs) are NOT considered
8: Designation test 4(3)(b)	Not "measures" but "other means" are considered	– Comparison of current benefits with benefits of other means – Disproportionality of costs of other means should be considered – Technical feasibility of other means should be considered
9: Designation	None	Not considered
10: Establishing MEP	All mitigation measures – that do not significantly adversely affect the specified uses or the wider environment; and – that ensure the best approximation to ecological continuum	– When assessing the adverse effects on the specified uses and on the wider environment, costs need to be considered – The benefits to the water body of applying the mitigation measures should be considered – Costs of mitigation measures (including disproportionality of costs) are NOT considered – Technical feasibility of mitigation measures NOT to be considered
11: Establishing GEP	Mitigation measures that – do not significantly adversely affect the specified uses or the wider environment; and – improve water body to slight deviation of MEP	– When assessing the adverse effects on the specified uses and on the wider environment, costs need to be considered – The benefits to the water body of applying the mitigation measures should be considered – Costs of mitigation measures (including disproportionality of costs) are NOT considered – Technical feasibility of mitigation measures NOT to be considered
For all water bodies (natural, Artificial and Heavily Modified Water Bodies):		
POM for reaching the environmental quality objectives (EQO)	All measures according to Article 11 WFD (including other means and mitigation measures considered in the designation process)	– Costs of measures (including disproportionality of costs) should be considered – Select the most cost-effective combination of measures to achieve the EQO – Technical feasibility of the measures should be considered

steps refer to the HMWB and AWB identification and designation process as shown in Fig. 2.1 (p. 12).

- The mitigation measures identified for achieving GEP were often considered as the basic measures of the Programme of Measures (POM). The link between GEP and measures was based on the present understanding of physical factors which were known or thought to influence the ecological status.
- Supplementary measures of the POM included among others measures targeted towards institutional instruments, measures promoting codes of good practice as well as the active involvement of stakeholders.
- In some cases, the difference between basic and supplementary measures (in POM) remained unclear. Therefore, the need was expressed for a clear definition and criteria for the two types of measures.

Chapter 11

Conclusions and Outlook

HMWB Case Studies as an Empirical Basis for Developing the CIS Guidance

The European Synthesis Project on Heavily Modified Water Bodies (HMWB) has been part of the activities of the WFD Working Group on HMWB of the Common Implementation Strategy (CIS). In the Synthesis Project, thirty-four case studies from eleven countries on the identification and designation of HMWB under the requirements of the WFD were evaluated. This evaluation included the identification of commonalities and differences in their approaches with regard to the identification and designation of HMWB and Artificial Water Bodies (AWB).

The research findings as presented in the different chapters of this book have formed a basis for the drafting, the revision and the production of the final CIS guidance document on the identification and designation of HWMB and AWB (WFD CIS Guidance Document No. 4 2003). The case studies have therefore been important in providing an empirical basis for the identification and designation process as well as practical input to the final guidance on HMWB and AWB and the accompanying toolbox. The experiences from the case studies increase the validity and the practical applicability of the final guidance document to most circumstances. It should be pointed out however that the methods and approaches used in the case studies, as presented in this book, do not always reflect the content of the final guidance document on HMWB and AWB.

The HMWB case studies have revealed the large variance of country-specific methods relevant to the identification and designation of HMWB and AWB. In this context, it has been a challenging and in parts difficult task to synthesise such large variance without overlooking the importance of each individual approach.

As mentioned in the beginning of this book, the HMWB case studies were carried out prior to the publication of the final CIS guidance document on HMWB and AWB. The HMWB identification and designation approach followed in the case studies was based on the draft process of Fig. 2.2 (p. 15), which was agreed by the CIS Working Group on HMWB during its initial activities. Some case studies, however, did not strictly follow this proposed draft identification and designation process but presented similar alternatives. The reason they used alternative approaches was to simplify the designation process as well as to target time and effort at complex cases of designation. Alternative approaches of the HMWB case studies have been described in more detail in Sect. 8.4 of this book.

Some elements of these alternative approaches have been considered for the production of the final CIS guidance document on HMWB and AWB. For instance, the authors of the UK-EW guideline report (Booker et al. 2002a) proposed a pre-screening methodology to screen out obviously Heavily Modified Water Bodies from obviously natural water bodies (see Fig. 5.1, p. 51). Pre-screening was followed by a screening step, which allowed targeting the designation effort at borderline HMWB designation cases (see Fig. 8.3, p. 131). Thus, cases that were obviously HMWB or obviously not HMWB were screened out early in the process. This approach was considered very useful for cases of large numbers of water bodies to be considered for designation. In the final guidance document on HMWB and AWB (WFD CIS Guidance Document No. 4 2003), a screening step similar to the UK approach has been included (see Step 3 in Fig. 2.1, p. 12). This screening step aims to screen out water bodies with only minor modifications and can be considered equivalent to the pre-screening step proposed by the UK-EW.

Some HMWB case studies also grouped water bodies with similar water types and hydromorphological changes in order to simplify the HMWB and AWB designation process. Such water bodies could then be tested for HMWB designation as a whole group. This approach has also been endorsed by the final CIS guidance document on HMWB and AWB (see Sect. 3 of the WFD CIS Guidance Document No. 4 2003).

Further Needs for Implementing the WFD

Additionally to contributing to the development of the guidance document and the toolbox, the HMWB case studies have revealed further guidance, policy and research needs for the purpose of implementing the WFD with regard to the identification and designation of HWMB and AWB.

Some of these issues were resolved in the final guidance document on HMWB and AWB (WFD CIS Guidance Document No. 4 2003) as discussed in the individual chapters of this book. Some issues, however, have remained partly unresolved.

Further guidance is required on the "best environmental practice" for modifications that support different types of uses. In some cases, such guidance exists (e.g. for flood defence) and needs to be collated and made available to all Member States, while for other uses guidance on the "best environmental practice" does not yet exist.

Further guidance is also needed on the types of measures that should be considered when defining the MEP and GEP. It is unclear to what extent measures in the catchment (and not only in the water body itself) should be taken into consideration. In relation to this, the authors of the case study on the Haringvliet Estuary (NL) requested a clarification of the role of contaminated sediment within the WFD. In the particular case study, it was not clear how essential "clean" sediment is for achieving GES, MEP or GEP or how new risks involved with new erosion of contaminated sediments should be dealt with.

From the evaluation of the HMWB case studies, further research needs have also been made apparent. Additional research should ensure that HMWB identification and designation decisions are taken with appropriate levels of confidence. Below, some recommendations are made that could be used for setting research priorities:

- There is a need to develop methods for the assessment of the structural status of rivers building on the experience of existing systems. Data on the morphological alterations of surface waters should be gathered.

- Research is needed on linking physical pressures to hydromorphology as well as linking hydromorphology to biological impacts. This includes the need for further diagnostic methodologies and frameworks that are required to separate impacts on biology resulting from different types of pressures (e.g. physical pressures versus water quality pressures). This is particularly the case when different pressures affect the same parameters (e.g. the non-physical modification impact of water abstraction versus the impact of a physical barrier on flow). Therefore, there is a need to adjust existing methods or to develop new methods that can distinguish between biological impacts resulting from different types of pressures.
- Further research is necessary on deriving exact cause-effect-relationships between restoration measures (to achieve GES) as well as mitigation measures (to achieve MEP and GEP) and the resulting ecological condition. At the moment, such estimations are accompanied by a high degree of uncertainty. In cases where poor ecological status results from both physical alterations and low water quality, the effect of restoration measures on water quality and the ecological status is nearly impossible to differentiate with current methodologies and knowledge. Moreover, it became apparent through the HMWB case studies that the difficulty in predicting the (quantitative) response of biology to specific mitigation measures made it impossible to appropriately define MEP.
- There is a need for methods of analysis to relate costs (which are generally economic in nature) with benefits to ecology. Due to the existing methodological difficulties, a systematic underestimation of such benefits occurs. The development and recognition of appropriate methodologies for quantifying the wider benefits of enhancing water body ecology should be a priority.

From the evaluation of the HMWB case studies, it additionally became apparent that there is a need to coordinate and integrate the WFD requirements for HMWB and AWB with other policy fields and other water uses. Divergent perceptions and requirements with regard to the importance of different (water) uses within a catchment may cause conflicts when restoration or mitigation measures are to be taken. Below, some cases of potential conflict with regard to spatial planning, energy policies, nature protection policies as well as the use of fish farming are briefly discussed as identified in the HMWB case studies.

Spatial Planning

The environmental objectives of the WFD as well as the designation of HMWB may cause conflicts with spatial planning policies in the Member States. This has been discussed by the authors of the case study on the Hagmolenbeek River in the Netherlands as illustrated below in Box 11.1.

Spatial planning aspects are not explicitly incorporated in the WFD. The above example shows that in terms of reaching the objectives of the WFD in an efficient and coordinated way, links between the WFD and spatial planning on a national level should be established.

Energy Policy

The authors of the case study on the Tummel River (UK) argued that energy policies in Member States should link to the objectives of the WFD and HMWB designation.

Box 11.1. HMWB Designation and Spatial Planning Policies in the Hagmolenbeek River, The Netherlands

According to authors of the case study on the Hagmolenbeek River (NL), agriculture was of major priority in the catchment of the Hagmolenbeek in terms of the Dutch spatial planning policy. This meant that the function nature and ecology had a lower priority than agriculture. This way of spatial planning policy and designation of functions conflicted with the WFD, according to which all waters should reach GES. Only if reaching GES were technically infeasible or if the costs were too high, lower ecological goals could be set. To solve this problem, the case study authors suggested that spatial planning aspects should be incorporated in the WFD. In this case study, the Dutch spatial planning policy also related to low feasibility with regard to the restoration of stream morphology. The space needed for restoration would have to be bought on a voluntary basis, leading to delays and incomplete restoration. If the spatial planning policy would designate nature as the most important function in the catchment, the water board would have more instruments to obtain the area along the stream.

Source: Case study on the Hagmolenbeek River, NL.

For instance, some ecological gains that may be set aside given the hydropower "protection" offered by HMWB designation could be achieved through future reductions in electricity consumption. This should fit well with policies such as those developed by the UK to increase both energy efficiency and the use of renewable sources; it would be logical to explicitly recognise that energy and water policies are linked.

According to the authors of the same case study, different solutions may be proposed among Member States in seeking to balance impacts on water body ecology against impacts on designated uses (in the case of hydropower, for instance, impacts on the use may then be linked to other environmental impacts such as emissions generated by other methods of electricity production). It should be therefore recognised that tensions between competing demands exist and clarification of the WFD requirements for Member States should be provided.

EU Nature Conservation Directives

HMWB designation and the application of necessary restoration measures to achieve GES might cause conflicts with the EU nature conservation directives. An example was given in the case study on Lake Veluwerandmeren in the Netherlands (see Box 11.2).

According to the final guidance document on HMWB and AWB (Sect. 6.4.3 of the WFD CIS Guidance Document No. 4 2003), the designation test Art. 4(3)(a) also has links to WFD Article 4(8) and 4(9) that require measures under the WFD to be consistent with the requirements of existing Community environmental legislation. For example, where the modified water body or its floodplain is (or is to be) designated under another directive such as the Fauna Flora Habitat or the Birds Directive, the requirements for these directives must be taken into account. "Restoration measures" that would result in conflicts with these directives should be considered as having a "significant effect on the environment" according to the designation text Art. 4(3)(a) (WFD CIS Guidance Document No. 4 2003).

Fish Farming

According to the authors of the case study on the Dalälven River (S), sustainable natural fish populations are one of the goals of the WFD. In this context, it is essential that the

Box 11.2. HMWB Designation and EU Nature Conservation Directives in Lake Veluwerandmeren, The Netherlands

According to the authors of the case study on Lake Veluwerandmeren, certain unnatural characteristics of Lake Veluwerandmenen stimulate the presence of birds. The unnatural water levels of the Veluwerandmeren, especially the low water levels in winter, lead to a higher density of birds, as they can more easily forage in shallow water. Secondly, eutrophication leads to a high biomass of fish, which is an important food source for the birds. Therefore, reduction of the nutrient load and the realisation of more natural water level dynamics to reach the objectives of the WFD would probably lead to a decrease in the diversity and abundance of birds, which is not acceptable according to the Birds Directive.

Source: Case study on Lake Veluwerandmeren, NL.

Box 11.3. The WFD and Fish Farming in the Dalälven River, Sweden

According to the authors of the case study on the Dalälven River, it can be expected that the resistance towards taking measures to meet the "river continuum concept" will not only be directly economical. In the case of Sweden, institutions with governmental connection like the Swedish Fish Health (Fiskhälsan) are primarily interested in prohibiting pests from fish in farms. This priority would most certainly discourage measures to facilitate fish migration in many river systems. This is due to the fact that increased migration possibilities will possibly increase the occurrence of pests, although pest outbreaks would probably not have such damaging effects on fish in nature as on fish in farms.

Source: Case study on the Dalälven River, S.

effects of fish farming should be judged with respect to the goals of the WFD so as not to counteract it. The case study authors recommended that costs for cultivating and releasing fish and the effects of those activities should be evaluated against costs for improving habitats to increase natural fish production as an alternative and more sustainable strategy. An example of potential conflict between the WFD and fish farming was presented in the case study on the Dalälven River (S) (see Box 11.3).

Based on the above examples, it is recommended to put the need for coordination and integration of HMWB requirements with other policy fields and water uses on the priority list. An implementation process of the WFD ignoring this need could very possibly compromise the achievement of the WFD's environmental objectives including those relevant for Artificial and Heavily Modified Water Bodies. The HMWB case studies showed the need for integration of policies and strategies on a national level but it may well be the case that integration efforts would be most useful also on the EU policy level.

Outlook

The HMWB case studies and the European Synthesis Project on Heavily Modified and Artificial Water Bodies were the beginning of a series of activities dealing with HMWB and AWB under the requirements of the WFD. The empirical value of the thirty-four case studies which is reflected in the final CIS guidance document on HMWB and AWB serves as reference and source of information for researchers and authorities involved in the WFD implementation in the different countries.

The final guidance on HMWB and AWB further is tested and validated along with other CIS guidance documents in the so-called pilot river basins across Europe during 2003 and 2004. The pilot river basins are part of the continuation of the WFD Common Implementation Strategy (CIS). The pilot basins have been chosen to test the practical application of the CIS guidance documents. In the light of further experiences gained during further testing and implementation, the guidance documents may be amended following the pilot testing exercises. The guidance documents hence remain "living documents" that will be adapted to further knowledge and testing results. As pointed out earlier in this chapter, further research and policy coordination is needed to reach well-informed decisions and maximum effectiveness of restoration and mitigation measures linked to Artificial and Heavily Modified Water Bodies. This book will serve as a solid basis in the pyramid of knowledge that will be accumulated in the coming years across Europe on the identification and designation of Artificial and Heavily Modified Water Bodies.

In 2005, the first reporting deadline to the European Commission is due, including information on the provisional identification of HMWB. It is interesting to view the quantity of water bodies that are identified as heavily modified, as well as the extent to which the approaches laid down in the final guidance document on HMWB and AWB is applied.

Chapter 12

Bibliography

12.1 Case Studies

Austria

Konecny R, Aschauer A, Chovanec A, Waringer J, Wimmer R, Waidbacher H, Schmutz S (2002a) Heavily Modified Waters in Europe – Case Study Danube. Federal Environment Agency, Vienna. Available online: http://www.sepa.org.uk/hmwbworkinggroup/

Konecny R, Aschauer A, Chovanec A, Wimmer R, Keckeis H (2002b) Heavily Modified Waters in Europe – Case Study Wienfluss. Federal Environment Agency, Vienna. Available online: http://www.sepa.org.uk/hmwbworkinggroup/

Konecny R, Aschauer A, Chovanec A, Wimmer R, Schmutz S (2002c) Heavily Modified Waters in Europe – Case Study Bregenzerach. Federal Environment Agency, Vienna. Available online: http://www.sepa.org.uk/hmwbworkinggroup/

Belgium

Vandaele K, De Bruyne I, Pauwels G, Willems I, Warmoes T (2002) Heavily Modified Waters in Europe – Case Study on the Dender River, the Mark River and Bellebeek River in Flanders. Soresma Environmental Consultants and Flemish Environmental Agency, Leuven and Antwerp. Available online: http://www.sepa.org.uk/hmwbworkinggroup/

Finland

Marttunen M, Hellsten S (2003) Heavily Modified Waters in Europe – A Case Study of Lake Kemijärvi, Finland. The Finnish Environment 630, Helsinki. ISBN 952-11-1417-7, ISBN 952-11-1418-5 (PDF), ISSN 1238-7312. Available online: http://www.environment.fi/publications

France

Agence de l'Eau Artois Picardie (2002) Heavily Modified Water Bodies – Case study on the River Authie, France. Available online: http://www.sepa.org.uk/hmwbworkinggroup/

Agence de l'Eau Rhin-Meuse (2002) Heavily Modified Water Bodies – Case Study on the River Sarre, France. Available online: http://www.sepa.org.uk/hmwbworkinggroup/

Agence de l'Eau Rhone Mediterranée Corse (2002) Heavily Modified Water Bodies – Case Study on the River Rhone, France. Available online: http://www.sepa.org.uk/hmwbworkinggroup/

Germany

Borchardt D, Podraza P (2002) Heavily Modified Waters in Europe – Case Study on the River Dhünn. Institute for Water Resources Research and Management, University Kassel, Kassel. Available online: http://www.sepa.org.uk/hmwbworkinggroup/

Frey M, Borchardt D, Funke M, Geffers K, Schleiter I (2002) Heavily Modified Waters in Europe – Case Study on the Elbe River. Institute for Water Resources Research and Management University Kassel, Kassel. Available online: http://www.sepa.org.uk/hmwbworkinggroup/

Funke M, Borchardt D, Frey M, Schleiter I, Geffers K (2002) Heavily Modified Waters in Europe – Case Study on the Seefelder Aach River. Institute for Water Resources Research and Management, University of Kassel, Kassel. Available online: http://www.sepa.org.uk/hmwbworkinggroup/

Podraza P, Greve K, Halle M, Zumbroich T, Müller A, Glacer D (2002a) Heavily Modified Waters in Europe – Case Study on the River Zwickauer Mulde. Buero fuer Umweltanalytik, Bonn and Essen. Available online: http://www.sepa.org.uk/hmwbworkinggroup/

Podraza P, Greve K, Halle M, Zumbroich T, Müller A, Glacer D (2002b) Heavily Modified Waters in Europe – Case Study on the River Ruhr. University of Essen, Institute of Ecology, Department of Hydrobiology, Essen. Available online: http://www.sepa.org.uk/hmwbworkinggroup/

Schleiter I, Borchardt D, Frey M, Funke M, Geffers K (2002) Heavily Modified Waters in Europe – Case Study on the River Lahn, Institute for Water Resources Research and Management, University Kassel, Kassel. Available online: http://www.sepa.org.uk/hmwbworkinggroup/

Greece

Paraskevopoulos A (2001) Heavily Modified Waters in Europe – Case Study on the River Nestos. Paraskevopoulos-Georgiadis EPE. Available online: http://www.sepa.org.uk/hmwbworkinggroup/

Netherlands

Backx JJGM, Berg GVD, Geilen N, de Hoog A, Houwing EJ, Ohm M, van Oirschot M, van Wijngaarden M (2002) Heavily Modified Waters in Europe – Case Study on the Haringvliet Estuary. RIZA, Dordrecht. Available online: http://www.sepa.org.uk/hmwbworkinggroup/

Lorenz CM (2001) Heavily Modified Waters in Europe – Case Study on the Hagmolen-Hegebeek Witteveen+Bos (W+B), Deventer. Available online: http://www.sepa.org.uk/hmwbworkinggroup/

Lorenz CM of W+B in association with DWR and RIVM (2001a) Heavily Modified Waters in Europe – Case Study on Lake Loosdrecht. Witteveen+Bos (W+B), DWR and RIVM, Deventer. Available online: http://www.sepa.org.uk/hmwbworkinggroup/

Lorenz CM of W+B in association with RDIJ and RIZA (2001b) Heavily Modified Waters in Europe – Case Study on the Veluwerandmeren. Witteveen+Bos (W+B), RDIJ and RIZA, Deventer. Available online: http://www.sepa.org.uk/hmwbworkinggroup/

Norway

Bjørtuft SK, Magnell JP, Koksvik JI (2002) Heavily Modified Waters in Europe – Case Study on the Beiarelva water course. Statkraft Grøner and Norwegian University of Science and Technology (NTNU), Lysaker and Trondheim. Available online: http://www.sepa.org.uk/hmwbworkinggroup/

Johansen SW, Magnell JP, Saltveit SJ, Sælthun NR (2002) Heavily Modified Waters in Europe – Case Study on the Suldalslågen River. Statkraft-Grøner, NIVA and LFI, Lysaker. Available online: http://www.sepa.org.uk/hmwbworkinggroup/

Spain

Diaz JA, Real M (2001) Heavily Modified Waters in Europe – Case Study on the river Lozoya (Tajo, Spain). Confederación Hidrográfica del Tajo, Calidad de Aguas and Limnos, S.A., Barcelona, Madrid. Available online: http://www.sepa.org.uk/hmwbworkinggroup/

Sweden

Beier U (2002) Heavily Modified Waters in Europe – Case Study on the River Dalälven. National Board of Fisheries, Institute of Freshwater Research, Drottingholm. Available online: http://www.sepa.org.uk/hmwbworkinggroup/

Jansson R (2002) Heavily Modified Waters in Europe – Case Study on the Umealven River in northern Sweden. Landscape Ecology Group, Department of Ecology and Environmental Science, Umeå University, Umeå. Available online: http://www.sepa.org.uk/hmwbworkinggroup/

Tullback K, Lindblad C (2001) Heavily Modified Waters in Europe – A Case Study on the Stockholm Archipelago, Baltic Sea. County Administrative Board of Stockholm, Environment and Planning Department and Department of Botany Stockholm University, Stockholm. Available online: http://www.sepa.org.uk/hmwbworkinggroup/

Weichelt AK (2001) Heavily Modified Waters in Europe – Case Study on the Emån River, Sweden. County Administrative Board Jönköping, Jönköping. Available online: http://www.sepa.org.uk/hmwbworkinggroup/

UK, Northern Ireland

Hale P, Corbelli D, Vincent C, Postle M, Fenn T, Ash J (2002) Heavily Modified Waters in Europe – Case Study on the River Lagan, the Tidal Lagan Transitional Water and the Port of Belfast Coastal Water, Northern Ireland. Environment and Heritage Service and Risk and Policy Analysts, Lisburn, London. Available online: http://www.sepa.org.uk/hmwbworkinggroup/

UK, England and Wales

Booker D, Dunbar M, Bass J, Dawson H, Gozlan R, Welton S, Latimer P, Rogerson H, Ash J, Fenn T, Postle M (2002a) Heavily Modified Waters in Europe – England and Wales Case Studies, Guidelines on identification, assessment and designation of rivers, Final Draft (Version 4). Submitted by the Environment Agency of England and Wales and the UK Government Department for Food, Environment and Rural Affairs, England and Wales

Booker D, Dunbar M, Folwell S, Stratford C, Latimer P, Rogerson H, Bass J, Dawson H, Gozlan R, Welton S, Ash J, Fenn T, Postle M (2002b) Heavily Modified Waters in Europe – Case Study on the River Kennet. Submitted by the Environment Agency of England and Wales and the UK Government Department for Food, Environment and Rural Affairs, England and Wales. Available online: http://www.sepa.org.uk/hmwbworkinggroup/

Booker D, Dunbar M, Stratford C, Latimer P, Rogerson H, Bass J, Dawson H, Gozlan R, Welton S, Ash J, Fenn T, Postle M (2002c) Heavily Modified Waters in Europe – Case Study on the Great Ouse Catchment. Submitted by the Environment Agency of England and Wales and the UK Government Department for Food, Environment and Rural Affairs, England and Wales. Available online: http://www.sepa.org.uk/hmwbworkinggroup/

Booker D, Dunbar M, Stratford C, Latimer P, Rogerson H, Bass J, Dawson H, Gozlan R, Welton S, Ash J, Fenn T, Postle M (2002d) Heavily Modified Waters in Europe – Case Study on the Tame Catchment. Submitted by the Environment Agency of England and Wales and the UK Government Department for Food, Environment and Rural Affairs, England and Wales. Available online: http://www.sepa.org.uk/hmwbworkinggroup/

Booker D, Dunbar M, Stratford C, Latimer P, Rogerson H, Bass J, Dawson H, Gozlan R, Welton S, Ash J, Fenn T, Postle M (2002e) Heavily Modified Waters in Europe – Case Study on the Sankey Catchment. Submitted by the Environment Agency of England and Wales and the UK Government Department for Food, Environment and Rural Affairs, England and Wales. Available online: http://www.sepa.org.uk/hmwbworkinggroup/

UK, Scotland

Black AR, Bragg OM, Duck RW, Findlay AM, Hanley ND, Morrocco SM, Reeves AD, Rowan JS (2002a) Heavily Modified Waters in Europe – Case Study on the River Tummel. Geography Department, University of Dundee and Department of Economics, University of Glasgow, Dundee, Glasgow. Available online: http://www.sepa.org.uk/hmwbworkinggroup/

Black AR, Bragg OM, Duck RW, Findlay AM, Hanley ND, Morrocco SM, Reeves AD, Rowan JS (2002b) Heavily Modified Waters in Europe – Case Study on the River Dee (Galloway, Scotland). Geography Department and Biological Sciences Institute, University of Dundee and Department of Economics, University of Glasgow, Dundee, Glasgow. Available online: http://www.sepa.org.uk/hmwbworkinggroup/

Black AR, Bragg OM, Caudwell CM, Duck RW, Findlay AM, Hanley ND, Morrocco SM, Reeves AD, Rowan JS (2002c) Heavily Modified Waters in Europe – Case Study on the Forth Estuary. Geography Department and Biological Sciences Institute, University of Dundee and Department of Economics, University of Glasgow, Dundee, Glasgow. Available online: http://www.sepa.org.uk/hmwbworkinggroup/

12.2 WFD-CIS Documents

WFD CIS Guidance Document No. 1 (2002) Economics and the Environment – The Implementation Challenge of the Water Framework Directive. August 2002. Published by the Directorate General Environment of the European Commission, Brussels, ISBN 92-894-4144-4, ISSN 1725-1087

WFD CIS Guidance Document No. 2 (2002) Identification of Water Bodies. December 2002. Published by the Directorate General Environment of the European Commission, Brussels, ISBN 92-894-5122-X, ISSN 1725-1087

WFD CIS Guidance Document No. 3 (2002) Analysis of Pressures and Impacts. December 2002. Published by the Directorate General Environment of the European Commission, Brussels, ISBN 92-894-5123-8, ISSN 1725-1087

WFD CIS Guidance Document No. 4 (2003) Identification and Designation of Artificial and Heavily Modified Water Bodies. January 2003. Published by the Directorate General Environment of the European Commission, Brussels, ISBN 92-894-5124-6, ISSN 1725-1087

WFD CIS Guidance Document No. 5 (2003) Transitional and Coastal Waters – Typology, Reference Conditions and Classification Systems. February 2003. Published by the Directorate General Environment of the European Commission, Brussels, ISBN 92-894-5125-4, ISSN 1725-1087

WFD CIS Guidance Document No. 10 (2003) Rivers and Lakes – Typology, Reference Conditions and Classification Systems. March 2003. Published by the Directorate General Environment of the European Commission, Brussels, ISSN 1725-1087

WFD CIS Working Group 2.2 on HMWB (2003a) Toolbox on identification and designation of Artificial and Heavily Modified Water Bodies, Final. 15 January 2003. Available online: http://forum.europa.eu.int/Public/irc/env/wfd/library

WFD CIS Working Group 2.2 on HMWB (2003b). Policy Summary to the final Guidance Document on Artificial and Heavily Modified Water Bodies, Final. 14 January 2003. Available online: http://forum.europa.eu.int/Public/irc/env/wfd/library

12.3 General Literature

Boon PJ, Fozzard IR, Holmes NTH, Maitland PS (2002) Developing a new version of SERCON (System for evaluating rivers for conservation). Aquat Conserv 12(4):439–455

Chave P (2001) The EU Water Framework Directive – An Introduction. IWA Publishing, London

Dynesius M, Nilsson C (1994) Fragmentation and flow. Regulation of river systems in the northern third of the world. Science 266:753–762

Grelsson G, Nilsson C (1982) Förändringar i en älvdals vegetation belysta genom upprepad fotografering. Svensk Botanisk Tidskrift 76:368–379

Hütte M (2000) Ökologie und Wasserbau – Ökologische Grundlagen von Gewässerverbauung und Wasserkraftnutzung. Parey-Buchverlag im Blackwell Wissenschafts-Verlag, Berlin and Wien

Länderarbeitsgemeinschaft Wasser (LAWA) (1998) Beurteilung der Wasserbeschaffenheit von Fließgewässern in der Bundesrepublik Deutschland – Chemische Gewässergüteklassifikation. LAWA-Arbeitskreis "Zielvorgaben" in Zusammenarbeit mit LAWA-Arbeitskreis "Qualitative Hydrologie der Fließgewässer", August 1998

Länderarbeitsgemeinschaft Wasser (LAWA) (2000) Gewässerstrukturgütekartierung in der Bundesrepublik Deutschland, Verfahren für kleine und mittelgroße Fließgewässer, Empfehlungen. Kulturbuch-Verlag GmbH, Berlin

Landesumweltamt Nordrhein-Westfalen (LUA NRW) (2001) Gewässerstrukturgüte in Nordrhein-Westfalen, Anleitung für die Kartierung mittelgroßer bis großer Fließgewässer. Landesumweltamt Nordrhein-Westfalen, Essen (LUA-Merkblätter 26)

Munné A, Prat, N Solà C (1998) QBR: Un índice rápido para la evaluación de la calidad de los ecosistemas de ribera. Tecnología del Agua 175:20–35

Munné A, Bonada N, Prat N, Rieradevall M, Solà C (2003) A simple field method for assessing the ecological quality of riparian habitat in rivers and streams: QBR index. Aquat Conserv 13:147–163

Poff NL, Allan JD, Bain MB, Karr JR, Restegaard KL, Richter BD, Sparks RE, Stromberg JC (1997) The natural flow regime: a paradigm for river conservation and restoration. Bioscience 47:769–784

Richter BD, Baumgartner JV, Braun DP, Powell J (1996) A method for assessing hydrological alteration within a river network. Conserv Biol 10:1163–1174

Richter BD, Baumgartner JV, Braun DP, Powell J (1998) A spatial assessment of hydrologic alteration within a river network. Regul River 14:329–340

Risk and Policy Analysts (RPA) (1998) The Environmental costs and benefits of water resources. Environment Agency, Bristol
Robinson M (1990) Impact of improved land drainage on river flows. Institute of Hydrology, Wallingford (Report 113)
Scheffer M (1998) The ecology of shallow lakes. Chapman & Hall, London
Schmutz S, Jungwirth M, Kaufmann M, Muhar S, Vogel B (2000) A multi-level concept for fish-based, river-type-specific assessment of ecolocical integrity. Hydrobiologia 422/423:279–289
Stanford JA, Frissell CA, Lichatowich JA, Liss WJ, och Coutant CC, Ward JV, Williams RN (1996) A general protocol for restoration of regulated rivers. Regul River 12:391–413
Turnpenny AWH (1999) Mitigation measures for migratory fish at hydro plants. Hydropower and Dams 5:83–88
Uhlmann D (1988) Hydrobiologie – Ein Grundriss für Ingenieure und Naturwissenschaftler, 3. Aufl. Gustav Fischer Verlag, Jena
UK Technical Advisory Group (2002) Draft paper on the definition of water bodies (TAG_019b). March 2002

Chapter 13

Annex I – Contributing Authors

Table 13.1. List of the HMWB working group members

Name	First name	Country	Organisation
Alvares	Teresa	Portugal	Ministerio do Ambiente e do Ordamento do Territorio
Andreadakis	Andreas	Greece	National Technical University
Aschauer	Arno	Austria	Umweltbundesamt Österreich
Aubert	Geraldine	France	Agence de l'Eau Artois-Picardie
Balashazy	Laszlo	Hungary	Ministry of Environment
Barkans	Idrikis	Latvia	Daugava Water Management
Barth	Friedrich	European Commission	Europäische Kommission, GD Umwelt
Bender	Michael		EEB/Grüne Liga
Bogie	Andrew	Ireland	Department of the Environment and Local Government
Bouma	Jestke	Netherlands	Institute for Inland Water Management and Waste Water Treatment (RIZA)
Bresser	Ton	Netherlands	National Institute of Public Health and Environmental Protection RIVM
Chovanec	Andreas	Austria	Umweltbundesamt Österreich
Christ	Andreas	Germany	Ministerium für Umwelt und Forsten
Coch Flotats	Antonio	Spain	Confederacion Hidrografica Ebro
Constantin	G.	Romania	Ministry of Water and Environmental Protection
Corbelli	David	UK	Scottish Environment Protection Agency (SEPA) Perth Office
Czerska	Bernadette	Poland	Ministry of the Environment
Demir	Tuncay		
Diaz Lazaro	Jose A.	Spain	Ministerio del Medio Ambiente
Dontchev	Vladimir	Bulgaria	Ministry of Environment
Duggan	Pat	Ireland	Department of the Environment and Local Government
Forrow	David	UK	Environment Agency of England and Wales (EA)
Ghini	Maria	Greece	Ministry of Development of Greece, Directorate of Water and Natural Resources
Grcar	Gabrijela	Slovenia	Ministry of Environment and Spatial Planning

Table 13.1. *Continued*

Name	First name	Country	Organisation
Hansen	Wenke	Germany	Ecologic – Institut für Internationale und Europäische Umweltpolitik
Heinonen	Pertti	Finland	Finnish Environment Institute
Hellsten	Seppo	Finland	Finnish Environment Institute, Hydrology and Water Management Division
Burakowska	Halina	Poland	Regional Water Management Board
Irmer	Ulrich	Germany	Umweltbundesamt, FG Binnengewässer
Janning	Jörg	Germany	Niedersächsisches Umweltministerium
Jarvi	Torbjorn	Sweden	National Board of Fisheries
Johansson	Caterina	Sweden	Swedish Environmental Protection Agency, Department of Environmental Assessment
Johansson	Daniel		EURELECTRIC
Kaczmarek	Bernard	Belgium	Bureau des Agences de l'Eau à Bruxelles
Kampa	Eleftheria	Germany	Ecologic – Institut für Internationale und Europäische Umweltpolitik
Kellet	Michael	UK	Scottish Executive Rural Affairs Department
Kinkor	Jaroslav	Czech Republic	Ministry of Environment
Kipgen	Robert	Luxemburg	Administration des Services Techniques de l'Agriculture
Kjellerup Larsen	Lars	Denmark	Ministry of the Environment, Danish Forest and Nature Agency
Koller-Kreimel	Veronika	Austria	Federal Ministry of Agriculture, Forestry, Environment and Water Management VII/1
Konecny	Robert	Austria	Umweltbundesamt Österreich
Kouvopoulos	Yannis	Greece	Public Power Corporation, Hydro-Electric Projects Development Department
Kraemer	Andreas	Germany	Ecologic – Institut für Internationale und Europäische Umweltpolitik
Kyrou	Kyriacos	Cyprus	Ministry of Agriculture, Natural Resources and Environment, Water Development Department
Lambot	Francis	Belgium	Ministry of the Region Wallone
Lazarou	Anastasia	Greece	Ministry of Environment, Physical Planning and Public Works
Ligtvoet	Willem	Netherlands	National Institute of Public Health and Environmental Protection RIVM
Makriyorgos	Charis	Greece	Public Power Corporation
Marcuello	Conchita	Spain	CEDEX
Marsden	Martin	UK	Scotland and Northern Ireland Forum for Environmental Protection (SNIFFER) and Scottish Environment Protection Agency (SEPA)
Martinet	Fabrice	France	Ministère de l'Aménagement du Territoire et de l'Environnement

Table 13.1. *Continued*

Name	First name	Country	Organisation
Marttunen	Mika	Finland	Finnish Environment Institute, Department for Expert Services
Moren-Abat	Marta-Cristina	European Commission	DG Environment B1 Water, Marine and Soil
Omorphos	Charis	Cyprus	Ministry of Agriculture
Ortiz-Casas	Jose Luis	Spain	Ministerio del Medio Ambiente
Pannonhalmi	Miklos	Hungary	North-Transdanubian District Water Authority
Pedersen	Tor Simon	Norway	Norwegian Water Researches and Energy Dir., Hydrology Department
Pilke	Ansa	Finland	Finnish Environment Institute
Puig	Alejandra	Spain	Ministerio del Medio Ambiente
Pio	Simone	Portugal	Ministerio do Ambiente e do Ordamento do Territorio
Puncochar	Pavel	Czech Republic	Ministry of Agriculture, Department of Water Management Policy
Rast	Georg	Germany	WWF
Rechenberg	Bettina	Germany	Umweltbundesamt
Rillaerts	Francis	Belgium	European Union of National Associations of Water Suppliers and Waste Water Services
Rivaud	Jean-Paul	France	Ministère de l'Aménagement du Territoire et de l'Environnement
Roelen	Ute	UK	WFD Economics WG
Scheuer	Stefan		EEB
Serban	Petru	Romania	National Administration Romanian Waters
Steiner	Anton	Germany	Bavarian Ministry of Environment
Tagg	Andrew	UK	EUREAU (Thames Water)
Tuurmann	Marko	Estonia	
Van Oirschot	Miel	Netherlands	Institute for Inland Water Management and Waste Water Treatment (RIZA)
Vanquaethem	Olivier	France	French Environment Ministry, Water Director
Van Riesen	Sigurd	Germany	European Water Association
Van Wijngaarden	Marjolein	Netherlands	Institute for Inland Water Management and Waste Water Treatment (RIZA)
Vincevicience	Violeta	Latvia	
Von Keitz	Stephan	Germany	Hessisches Ministerium für Umwelt, Landwirtschaft und Forsten
Warmoes	Thierry	Belgium	Flemish Environment Agency
Walcykiewicz	Tomasz	Poland	Ministry of Environment, Department of Water Resources
Zavadsky	Ivan	Slovakian Republic	Ministry of Environment

Table 13.2. List of the HMWB case study authors; authors have been listed alphabetically for each country

Country	Names of authors
Austria	Arno, Aschauer
	Chovanec, Andreas
	Keckeis, Hubert
	Konecny, Robert
	Schmutz, Stefan
	Waringer, Johann
	Wimmer, Reinhard
Belgium	De Bruyne, Ingrid
	Pauwels, Gert
	Vandaele, Karel
	Warmoes, Thierry
	Willems, Isabelle
Germany	Borchardt, Dietrich
	Frey, Michaela
	Funke, Markus
	Glacer, Dirk
	Halle, Martin
	Kuellmar, Ingrid
	Podraza, Petra
	Müller, Andreas
	Zumbroich, Thomas
Spain	Diaz, Jose-Antonio
	Real, Montserrat
Finland	Hellsten, Seppo
	Marttunen, Mika
France	Aubert, Geraldine
	Demortier, Guillaume
	Stroffek, Stéphane
Greece	Paraskevopoulos, Alexis
The Netherlands	Backx, J. J. G. M.
	Berg, G. v. d.
	Geilen, N.
	Hoog, A. de
	Houwing, E. J.
	Lorenz, C. M.
	Ohm, M.
	Oirschot, M. van
	Wijngaarden, M. van
Norway	Bjørtuft, Sigurd K.
	Johansen, Stein W.
	Koksvik, Jan Ivar
	Magnell, Jan-Petter
	Saltveit, Svein Jakob
	Saelthun, Nils Roar
Sweden	Beier, Ulrike
	Jansson, Roland
	Lindblad, Cecilia
	Tullback, Klara
	Weichelt, Ann-Karin
United Kingdom (EW)	Ash, John
	Bass, Jonathan
	Booker, Douglas
	Dawson, Hugh
	Dunbar, Michael
	Fenn, Teresa
	Gozlan, Rodolphe
	Latimer, Peter
	Postle, Meg
	Rogerson, Helen
	Stratford, Charlie
	Welton, Stewart
United Kingdom (S)	Black, A. R.
	Bragg, O. M.
	Caudwell, C. M.
	Duck, R. W.
	Findlay, A. M.
	Hanley, N. D.
	Morrocco, S. M.
	Rowan, J. S.
	Reeves, A. D.
United Kingdom (NI)	Ash, John
	Corbelli, David
	Hale, Peter
	Postle, Meg
	Venn, Teresa
	Vincent, Claire

Table 13.3. List of HMWB case study contacts

Country	Name of case study	Name	Institution	E-mail	Telephone
A	Bregenzerach R.	Konecny, Robert	Umweltbundesamt Österreich	Konecny@ubavie.gv.at	(43)1313043581
	Danube R.	Konecny, Robert	Umweltbundesamt Österreich	Konecny@ubavie.gv.at	(43)1313043581
	Wienfluss	Konecny, Robert	Umweltbundesamt Österreich	Konecny@ubavie.gv.at	(43)1313043581
B	Dender R.	Vandaele, Karel	SORESMA	Karel.vandaele@soresma.be	(32) 3-2215540
D	Elbe R.	Frey, Michaela	University of Kassel	m.frey@bauing.uni-kassel.de	(49) 561-8043949
	Seefelder Aach R.	Funke, Markus	University of Kassel	Funkem@hrz.uni-kassel.de	(49) 561-8043912
	Lahn R.	Kuellmar, Ingrid	University of Kassel	Ingrid.kuellmar@uni-kassel.de	(49) 561-8043991
	Ruhr R.	Podraza, Petra	University of Essen	Petra.podraza@uni-essen.de	(49) 201-1833868
	Mulde R.	Podraza, Petra	University of Essen	Petra.podraza@uni-essen.de	(49) 201-1833868
	Dhünn R.	Borchardt, Dietrich	University of Essen	Dietrich.Borchardt@uni-kassel.de	(49) 561-8043912
E	Lozoya R.	Diaz, Jose-Antonio	Ministerio del Medio Ambiente	Joseantonio.diaz@chtajo.es	(34) 91-5350500
SF	Kemijärvi L.	Marttunen, Mika	Finnsh Environment Institute	Mika.marttunen@ymparisto.fi	(358) 9-403000
F	Authie R.	Aubert, Geraldine	Agence de l´Eau Artois-Picardie	G.Aubert@eau-artois-picardie.fr	
	Sarre R.	Demortier, Guillaume	Agence de l´Eau Rhin-Meuse	DEMORTIER.G@Eau-Rhin-Meuse.fr	(33) 3-87344841
	Rhone R.	Stroffek, Stéphane	Agence de l´Eau Rhone-Mediterranée-Corse	Stephane.STROFFEK@eaurmc.fr	
GR	Nestos R.	Kouvopoulos, Yannis	Public Power Corporation	Tsmys3@daye.gr	
NL	Haringvliet Est.	Backx, J. J. G. M.	RIZA	J.Backx@riza.rws.minvenw.nl	(31) 78-6332736
	Hagmolenbeek-Hegebeek R.	Lorenz, C. M.	Witteveen and Bos	c.lorenz@witbo.nl	(31) 570-697272
	Loosdrecht L.	Lorenz, C. M.	Witteveen and Bos	c.lorenz@witbo.nl	(31) 570-697272
	Veluwerandmeren	Lorenz, C. M.	Witteveen and Bos	c.lorenz@witbo.nl	(31) 570-697272

Table 13.3. *Continued*

Country	Name of case study	Name	Institution	E-mail	Telephone
NO	Suldalslagen R.	Pedersen, Tor Simon	Norwegian Water Researches and Energy Dir/Hydrology Dept	tsp@nve.no	(47) 22-959205
	Beiarn R.	Bjørtuft, Sigurd K.,	Statkraft Grøner as	skb@statkraftgroner.no	
S	Eman R.	Weichelt, Ann-Karin	County Administrative Board Jönköping	Lansstyrelsen@f.lst.se	(46) 36-395000
	Daläven R.	Beier, Ulrike	National Board of Fisheries, Institute of Freshwater Research	Ulrika.Beier@fiskriverket.se	(46) 8- 590338
	Umealven R.	Jansson, Roland	Swedish Environmental Protection Agency/Department of Environmental Assessment	Roland@eg.umu.se	(46) 90-7869573
	Archipelago, Baltic Sea	Tullback, Klara	County Administrative Board	Klara.tullback@ab.lst.se	(46) 8-7854103
UK (EW)	Kennet R. (Thames)	Dunbar, Michael	Centre for Ecology and Hydrology	Mdu@ceh.ac.uk	(44) 1491-838800
	Tame R.	Dunbar, Michael	Centre for Ecology and Hydrology	Mdu@ceh.ac.uk	(44) 1491-838800
	Sankey Brook	Dunbar, Michael	Centre for Ecology and Hydrology	Mdu@ceh.ac.uk	(44) 1491-838800
	Great Ouse R.	Dunbar, Michael	Centre for Ecology and Hydrology	Mdu@ceh.ac.uk	(44) 1491-838800
UK (S)	Forth Estuary	Black, A. R.	Geography Department, University of Dundee	a.z.black@dundee.ac.uk	(44) 1382-344434
	Tummel R.	Black, A. R.	Geography Department, University of Dundee	a.z.black@dundee.ac.uk	(44) 1382-344434
	Dee R.	Black, A. R.	Geography Department, University of Dundee	a.z.black@dundee.ac.uk	(44) 1382-344434
UK (NI)	Lagan R.	Corbelli, David	SEPA	David.corbelli@sepa.org.uk	(44) 17-86457700

Chapter 14

Annex II – Case Study Evaluation Tables and Methods

Table 14.1. Water body categories and subgroups of the HMWB case studies

Country	Name of case study	River	Lake	Transitional waters	Coastal waters	Navigation subgroup (lead: D)	Hydropower subgroup (lead: A)
A	Bregenzerach R.	+					+
	Danube R.	+				+	+
	Wienfluss R.	+					
B	Dender R.	+				+	
D	Elbe R.	+				+	
	Seefelder Aach R.	+					+
	Lahn R.	+				+	+
	Ruhr R.	+					+
	Mulde R.	+					
	Dhünn R.	+					
E	Lozoya R.	+					+
SF	Kemijärvi L.		+				+
F	Authie R.	+					
	Sarre R.	+					
	Rhone R.	+					
GR	Nestos R.	+					+
NL	Haringvliet Est.			+		+	
	Haqmolenbeek-Hegebeek R	+					
	Loosdrecht L.		+				
	Veluwerandmeren		+				
NO	Suldalslagen R.	+					+
	Beiarn R.	+					+
S	Emån R.	+					+
	Dalaven R.	+					+
	Umealven R.	+					+
	Archipelago, Baltic Sea				+		
UK (EW)	Kennet R. (Thames)	+				+	
	Tame R.	+					
	Sankey Brook	+				+	
	Great Ouse R.	+				+	
UK (Scot)	Forth Est.			+			
	Tummel R.	+					+
	Dee R.	+					+
UK (NI)	Lagan R.	+				+	

Table 14.2. Catchment area and length/size of the case study areas. *Source:* Extracted from the case studies by the authors

Country	Name of case study	Small catchment	Medium catchment	Large catchment	Very large catchment	Length/size	Mean depth (or other significant parameter)
Unit		km^2 (10–100)	km^2 (100–1 000)	km^2 (1 000–10 000)	km^2 (>10 000)	km/km^2	
A	Wienfluss River	+				34 km	
	Bregenzerach River		+			68 km	
	Danube River				+	2 850 km (studied section: 50 km)	
B	Dender River			+		70 km	
D	Elbe River				+	1 091 km (730 km within Germany)	
	Seefelder Aach River		+			54 km	
	Lahn River			+		242 km	
	Ruhr River			+		219 km	
	Mulde River			+		167 km	
	Dhünn River		+			39.6 km	Reservoir: 440 ha, 53 m depth
E	Lozoya River		+			77.7 km	
SF	Kemijärvi Lake				+	128–285 km^2 (depending on water level)	Mean depth 1.7–4.5 m
F	Authie River		+			95 km	
	Sarre River		+			130 km	
	Rhone River				+	522 km in France (total length = 812 km)	
GR	Nestos River			+		234 km	
NL	Haringvliet Estuary				+	65 km	
	Hagmolenbeek-Hegebeek River	+				27 km	

Table 14.2. *Continued*

Country Unit	Name of case study	Small catchment km^2 (10–100)	Medium catchment km^2 (100–1 000)	Large catchment km^2 (1 000–10 000)	Very large catchment km^2 (>10 000)	Length/size km/km^2	Mean depth (or other significant parameter)
	Loosdrecht Lake	+				10 km^2	Depth: 1.85 m
	Veluwerandmeren (lake_		+			Nuldernauw: 6.64 km^2; Wolderwijd: 18.25 km^2; Veluwemeer: 30.22 km^2; Drontermeer 4.76 km^2	Depth: 1.30–2.00 m
NO	Suldalslagen River			+		22 km (case study)	
	Beiarn River		+			50 km	
S	Emån River			+		240 km (Emån), studied section is 23 km (Solgenan)	
	Dalälven River				+	555 km	Mean discharge: 353 m^3 s^{-1}
	Umealven River				+	470 km	
	Baltic sea coastal areas					Study areas: 1 km–3.8 km^2	Depth between 10–30 m
UK (EW)	Thames (Kennet and Avon)			+		Ca. 70 km (Kennet, main channel only – this does not include tributaries)	
	Tame River		+			140 km	
	Sankey Brook		+			126 km	
	Great Ouse River			+		1 146 km	
UK-S	Forth Estuary			+		93.9 km^2	Depth: <20 m
	Tummel River			+		1 253 km river channel	
	Dee River			+		78.5 km (588 km river channel)	
UK-NI	Lagan River		+			70 km	

Table 14.3. Case study results on the identification of water bodies (WBs). *Source:* Extracted from the case studies by the authors

Country	Name of case study	Identified water bodies (WBs) – verbal description of WBs	Groups of water bodies	Main criteria for identification (numbers in parenthesis indicate order of use of criteria)
A	Wienfluss R.	2 WBs Upper WB with minor impacts/downstream WB heavily impacted by flood protection		Effective management units
	Bregenzerach R.	4 WBs Natural/residual (2 WBs)/hydropeaking reach		Effective management unit
	Danube R.	1 WB		Effective management unit
B	Dender R.	18 WBs		Hydromorphology
D	Elbe R.	3 WBs Up/middle/downstream		Physical features, 'natural conditions'
	Seefelder Aach R.	1 WB		Effective management unit
	Lahn R.	5 WBs Up (2 WBs)/middle (2 WBs)/downstream		Hydromorphology/effective management unit
	Ruhr R.	3 management units upper/middle/lower Part of the lower unit is the core region of the study and consists of 2 WBs ; WB 1: mainly free flowing; WB 2: mainly impounded (each WB of the lower unit consists of several subbodies)		Effective management units (1st)/WFD typology system A (2nd)
	Mulde R.	7 WBs		Physical features (1st)/effective management units (2nd)
	Dhünn R.	1 WB		No criteria mentioned/no distinct WBs
E	Lozoya R.	12 WBs	3 groups: – WB Group 1 (natural streams) (4 WBs) – WB Group 2 (reservoirs) (5 WBs) – WB Group 3 (regulated river) (1 WB) – WB Group 3b (final regulated river) (2 WBs)	Effective management unit

Table 14.3. *Continued*

Country	Name of case study	Identified water bodies (WBs) – verbal description of WBs	Groups of water bodies	Main criteria for identification (numbers in parenthesis indicate order of use of criteria)
SF	Kemijärvi L.	7 WBs	2 groups: – main lake area and smaller sub-basins (1 WB) – permanently detached lakes (6 WBs)	Effective management unit/hydromorphology
	Sub-case River Raudanjoki	7 WBs		
	River Kemijoki (see Box 4.1)	*Alternative 1:* 1 WB *Alternative 2:* 2 WBs *Alternative 3:* 7 WBs (each impoundment is 1 WB)		Effective management units
F	Authie R.	(No division into different WBs; referrence to low, central and high valley)		No identification criteria
	Sarre R.	28 WBs (referred to as sections)		WFD typology (1st)/hydromorphology (most significant tributaries in terms of flow rate) (2nd)/effective management unit (3rd)
	Rhone R.	15 WBs	3 groups: – WBs heavily modified (6 WBs) – WBs created (artificial) (7 WBs) – WBs not affected by hydromorphological changes (2 WBs)	Hydromorphology and physical features (valley shape, hydrological patterns)/effective management units (infrastructure, impacts)
GR	Nestos R.	4 WBs Upstream of reservoir/reservoir/down-stream of reservoir/delta		Effective management units (1st)/typology system A (2nd)
NL	Haringvliet Est.	3 WBs Freshwater, mixed, tidal		Hydromorphology, supported by biological conditions (1st)/effective management units (2nd)
	Hagmolenbeek-Hegebeek R.	2 WBs Up/downstream		Hydromorphology/effective management unit
	Loosdrecht L.	1 WB		Hydromorphology/biological condition
	Veluwerandmeren	1 WB		Hydromorphology/biological conditions

Table 14.3. *Continued*

Country	Name of case study	Identified water bodies (WBs) – verbal description of WBs	Groups of water bodies	Main criteria for identification (numbers in parenthesis indicate order of use of criteria)
NO	Suldalslagen R.	1 WB		Effective management unit
	Beiarn R.	4 WBs Unaffected upstream/intake with affected flow/reduced flow (2 WBs)		Effective management units (1st) hydromorphology/WFD typology system A (2nd)
S	Emån R.	4 WBs		Biological status (1st)/typology system A (2nd)
	Dalälven R.	WB 1 (reservoir + downstream reach) WB 2 and 3 (regulated lakes) WB 4 and 5 (dams and HP plants)		Biological conditions (ecologically relevant scale)/effective management unit (types of constructions and impacts)
	Umealven R.	18 WBs Reservoirs (5 WBs)/run-off river impoundment (12 WBs)/each downstream of dam (2 WBs)	2 groups (according to ecoregions): – borealic ecoregion (5 WBs) – fennoscandian ecoregion (13 WBs)	WFD typology system A (1st)/physical features (2nd)/effective management unit (3rd)
	Baltic coast	3 different study areas		
UK (EW)	Kennet R.	8 WBs	4 groups: – heavily urbanised canal sections – semiurban canal sections – rural – rural canal sections	Effective management unit (based on pressures and physical alterations)
	Tame R.	3 WBs (7 stretches which could later be distinct WBs)		Hydromorphology/effective management unit
	Sankey Brook	17 WBs	3 groups (according to use): – urban (6 WBs) – semi-urban (4 WBs) – rural (7 WBs)	Water quality stretches (1st)/hydromorphology and effective management unit (2nd)
	Great Ouse R.	Subcase Wissey river: 3 WBs Subcase Great Ouse: 1 WB		Water quality stretches (1st)/hydromorphology and effective management unit (2nd)

Table 14.3. *Continued*

Country	Name of case study	Identified water bodies (WBs) – verbal description of WBs	Groups of water bodies	Main criteria for identification (numbers in parenthesis indicate order of use of criteria)
UK (S)	Forth Est.	3 WBs		Effective management units backed by physical conditions
	Tummel R.	57 WBs	7 groups (on the basis of hydropower stations)	Hydromorphology (1st)/effective management unit (2nd)/hydromorphology (subdivisions) (3rd)
	Dee R.	42 WBs	7 groups (on the basis of hydropower stations)	Hydromorphology (1st)/effective management unit (2nd)/hydromorphology (subdivisions) (3rd)
UK (NI)	Tummel R.	5 Wbs Up/middle/downstream, inner/outer harbour		Physical features (1st)/effective management unit (2nd)

Table 14.4. Specified use "navigation", its related physical alterations and changes of hydromorphology. *Source:* Extracted from the case studies by the authors

Specified use	Physical alteration[a]	Changes of hydromorphology
Navigation	*Rivers: Channels/beds*	*Rivers: Hydrological changes/impacts*
	Locks, mills	Disruption of river continuum
	Dams and weirs	Change of hydrological regime
	Sluices	Change from running river to slowly flowing river
	Dredging operations/channel maintenance	Change of hydraulic characteristics
	Straightening	Reduced velocity/low flow above weirs
	Canalisation	Water level increase behind weir
	River widening/deepening	Low water level downstream of weir
	Reservoirs/impoundment	Change of groundwater level in the area of weirs (dropping below and rising above weirs)
	Cutting of meanders and wetlands	Reduced flow in riverbed
	Change in channel course	Elevated flow velocity due to channel regulation
	Removal of islands and river divisions	Decrease of discharge fluctuations (min. and max.)
	Fixation of riverbed	Change of storage capacity
	Rivers: Banks/riparian zones	Disconnection from former catchment and discharge into a canal
	Bank reinforcement/fixation	*Rivers: Morphological changes/impacts*
	Dikes	Loss of morphological variety in cross and longitudinal profile
	Culverts	Change of linear profile
	Floating quays	Loss of original length of river
	Moorings/marinas	Decreased sinuosity
		Change of river depth/width
		Disruption of sediment transport
		Disturbance of bed load
		Reduced riverbed diversity
		Change of particle size of the riverbed (from gravel to silt)
		Increased sedimentation behind weir
		Change of structural characteristics
		Uniform bank structure
		Detachment/restriction of natural floodplains
		Reduced exchange between river and riparian zone
		Loss of riparian zone
		Silting of riparian zone
		Detached/lost oxbow lakes/wetlands
	Lakes: Basins/beds	*Lakes: Hydrological changes/impacts*
	Creation of islands (use of sand from shipping channel)	Fixed water level/water level management (high level in summer, low level in winter)
	Shipping channels	*Lakes: Morphological changes/impacts*
	Deepening of shallow areas	Unnatural lake morphology

Table 14.4. *Continued*

Specified use	Physical alteration[a]	Changes of hydromorphology
Navigation	*Coastal waters: Substratum*	*Coastal waters: Hydrological changes/impacts*
	Shipping channels	Change in flow direction
	Deepening of shallow areas	Interference with water replacement due to quays
	Channels/waterways	*Coastal waters: Morphological changes/impacts*
	Coastal waters: Shorelines	Bottom erosion
	Landing stages/jetties/quays	Resuspension of silt from boats
	Harbours	Shore erosion
	Transitional waters: Shorelines	*Transitional waters:* Hydrological changes/impacts
	Shoreline reinforcement/sea walls	No changes explicitly noted in the case studies
	Harbours/ports	*Transitional waters:* Morphological changes/impacts
	Transitional waters: Channels/beds	Interruption of sediment exchange with intertidal areas
	Shipping channels	Altered dynamics of sediment transport throughout the estuary

[a] The physical alterations and hydromorphological changes in the same row of the table do not necessarily or exclusively correspond to one another. The tables should be read vertically rather than horizontally.

Table 14.5. Specified use "flood protection", its related physical alterations and changes of hydromorphology. *Source:* Extracted from the case studies by the authors

Specified use	Physical alteration[a]	Changes of hydromorphology
Flood protection	*Rivers: Channels/beds*	*Rivers: Hydrological changes/impacts*
	Flood protection dams	Reduced fluvial dynamics
	Straightening	Disruption of river continuum
	Cutting of meanders and wetlands	Decrease of discharge fluctuations (min. and max.)
	Dredging/channel maintenance	Accelerated flood waves
	Storage reservoirs/impoundment	Loss of vital periodic flooding
	Complete re-location of river channel	Elevated flow velocity (channel regulation)
	Channelisation	Change from running river to slowly flowing river
	Weirs	Reduced flooding volume
	Change of river depth	Anthropogenic peak flows
	Rivers: Banks/riparian zones	Change of groundwater level in the area of weirs (dropping below and rising above weirs)
	Bank reinforcement/fixation	Reduced velocity (above weirs)
	Artificial channels (parallel subsidiary)	Water level increase behind weir
	Embankment	Low water level downstream of weir
	Dikes	Reduced flow in riverbed
	Culverts	Creation of impounded lakes
		Disconnection from former catchment and discharge into a canal
		Rivers: Morphological changes/impacts
		Loss of morphological variety in cross and longitudinal profile
		Change of river depth/width
		Uniform bank structure
		Steep bank profile
		Disturbance of bed load
		Loss of original length of river
		Decreased sinuosity
		Increased sedimentation behind weir
		Disruption of sediment transport
		Lowered river bottom downstream of weir
		Reduced riverbed diversity
		Change of particle size of the riverbed (from gravel to silt)
		Reduced diversity of bank structure
		Detachment/restriction of natural floodplains
		Reduced longitudinal and lateral connectivity
		Loss of riparian zone
		Silting up of riparian zone
		Detached/lost oxbow lakes/wetlands

Table 14.5. *Continued*

Specified use	Physical alteration[a]	Changes of hydromorphology
Flood protection	*Lakes: Shorelines*	*Lakes: Hydrological changes/impacts*
	Bank reinforcement/fixation	Fixed water level/water level management (high level in summer, low level in winter)
		Lakes: Morphological changes/impacts
		Unnatural lake morphology
	Coastal waters: Shorelines	*Coastal waters: Hydrological changes/impacts*
	Straightening of shoreline	Change in flow direction
	Sea walls	Interference with water replacement due to quays
		Coastal waters: Morphological changes/impacts
		Bottom and shore erosion
	Transitional waters: Shorelines	*Transitional waters: Hydrological changes/impacts*
	Shoreline reinforcement/sea walls	Altered tidal prism due to land claim
		Reduced tidal volume
		Reduced tidal amplitude
		Transitional waters: Morphological changes/impacts
		Interruption of sediment exchange with intertidal areas
		Altered dynamics of sediment transport throughout the estuary
		Reduced intertidal area

[a] The physical alterations and hydromorphological changes in the same row of the table do not necessarily or exclusively correspond to one another. The tables should be read vertically rather than horizontally.

Table 14.6. Specified uses "hydropower/water supply", its related physical alterations and changes of hydromorphology. *Source:* Extracted from the case studies by the authors

Specified use	Physical alteration[a]	Changes of hydromorphology
Hydropower/ water supply	*Rivers: Channels/beds*	*Rivers: Hydrological changes/impacts*
	Hydropower stations/dams	Disruption of river continuum
	Storage reservoirs/impoundment	Artificial flow regime/discharge (upstream and downstream)
	Run-off-river structures	Change from river or estuary to freshwater lake
	Catchwater intakes	Reduced discharge/flow and drying out (downstream of dams and catchwaters)
	Water diverting structures	Reduced water velocity
	Pipelines and aqueducts	Reduced flow in riverbed
	Weirs	Violation of natural seasonal in flow regime
	Straightening	Extreme peaking amplitudes
	Channelisation	Reduced flood peaks
	Clearings (removal of boulders and stones, deepening)	Reduced occurrence of large floods
	Rivers: Banks/riparian zones	Loss of bed scouring floods
	Bank reinforcement/fixation	Seasonal changes in flow regime
	Removal of riparian forest	Cross-catchment transfer
		Change in total water supply to segments of the river
		Permanent diversion/zero discharge downstream
		Reduction of glacier coverage/change in annual flow distribution (lower in summer)
		Increased ice cover upstream of dams
		Disappearance of ice jam events
		Rivers: Morphological changes/impacts
		Linear profile modified/change in river profile
		Change in area and perimeter of reservoir
		Disruption/reduction of sediment transport
		Change in streambed morphology
		Reduced riverbed diversity
		Accumulation of sediment in reservoir bottom/before dam
		Altered channel morphology downstream of dam
		Accumulation of sediments in upper river parts/reduced river transport capacity
		Reduction of total load of suspended material
		Interruption of litter transport
		Change from diverse substrate to fine particle sediment (siltation)
		Inadequate supply of sediment to delta

Table 14.6. *Continued*

Specified use	Physical alteration[a]	Changes of hydromorphology
Hydropower/ water supply		*Rivers: Morphological changes/impacts (continued)*
		Erosion downstream of dam
		Increased riverbed level
		Reduced exchange between river and riparian zone
		Reduced diversity of bank structure
		Detachment/restriction of natural floodplains
		Restriction of riparian zone
		Detached oxbow lakes/wetlands
	Lakes: Basins/beds	*Lakes: Hydrological changes/impacts*
	Hydropower stations/dams	Lake/reservoir level regulation
	Storage reservoirs/impoundment	Increase of lake/reservoir level fluctuation
	Lakes: Shorelines	Low flow/turn over of water (in detached lakes)
	Bank reinforcement/fixation	Raise of natural lake level (especially in summer)
	Removal of riparian forest	Decrease of water level during ice-covered period – large areas under ice exposure
		Lakes: Morphological changes/impacts
		Erosion of lake margins/geomorphologic change of littoral zone
		Change in riparian vegetation/bank morphology

[a] The physical alterations and hydromorphological changes in the same row of the table do not necessarily or exclusively correspond to one another. The tables should be read vertically rather than horizontally.

Table 14.7. Specified uses "agriculture/forestry", its related physical alterations and changes of hydromorphology. *Source:* Extracted from the case studies by the authors

Specified use	Physical alteration[a]	Changes of hydromorphology
Agriculture/ forestry	*Rivers: Channels/beds*	*Rivers: Hydrological changes/impacts*
	Timber floating structures	Change of hydrological regime
	Diversion dams/weirs	Change of hydraulic characteristics
	Straightening	Artificial flow
	(Water abstraction/intakes)	Affected connection with groundwater
	Rivers: Banks/riparian zones	Lower groundwater buffer capacity (due to drainage)
	Land drainage/ditches	Large seasonal variations in flow quantity
	Artificial channels (parallel subsidiary)	*Rivers: Morphological changes/impacts*
	Land reclamation	Sediment transport into stream from field erosion
	Land use change	Change of erosion/sediment pattern (sediment supply)
	Soil and bank erosion	Change/reduction of structural characteristics
	Planting of water-demanding trees (poplar)	Restriction of natural floodplains
	Cattle fences	Change in river profile
	Removal of riparian forest	Detachment of wetland
	Dikes	
	Lakes: Banks	*Lakes: Hydrological changes/impacts*
	Land reclamation	Reduction of seepage water into lake
		Lakes: Morphological changes/impacts
		No changes explicitly noted in the case studies
		Other impacts
		Change of water quality (input of run-off)

[a] The physical alterations and hydromorphological changes in the same row of the table do not necessarily or exclusively correspond to one another. The tables should be read vertically rather than horizontally.

Table 14.8. Specified uses "urbanisation/industry", its related physical alterations and changes of hydromorphology. *Source:* Extracted from the case studies by the authors

Specified use	Physical alteration[a]	Changes of hydromorphology
Urbanisation/ industry	*Rivers: Channels/beds*	*Rivers: Hydrological changes/impacts*
	Dams and weirs	Change of hydrological regime
	Straightening	Change of hydraulic characteristics
	Gravel extraction/spoil grounds	Altered frequency of flooding
	Fixation of riverbed	'Flashier' flow due to reduced infiltration
	Pits/mines/slagheaps	*Rivers: Morphological changes/impacts*
	(Water abstraction/intakes)	Change of erosion/sediment pattern (sediment supply)
	Rivers: Banks/riparian zones	Change of structural characteristics (river depth, width)
	Drainage/ditches	Uniform bank structure
	Channels/waterways	Reduced exchange between river and riparian zone
	Bank reinforcement	Restriction of natural floodplains
	Land reclamation	Loss of riparian zone
	Infrastructure (buildings, roads, bridges)	Reduced riverbed diversity
		Decreased sinuosity
	Lakes: Basins/beds	*Lakes: Hydrological changes/impacts*
	Sand extraction	No changes explicitly noted in the case studies
		Lakes: Morphological changes/impacts
		Creation of holes in lake bottom (sand extraction)
	Transitional waters: Shorelines	*Transitional waters: Hydrological changes/impacts*
	Land reclamation	Altered tidal prism due to land claim
		Transitional waters: Morphological changes/impacts
		No changes explicitly noted in the case studies
		Other impacts
		Change of water quality (input of wastewater)

[a] The physical alterations and hydromorphological changes in the same row of the table do not necessarily or exclusively correspond to one another. The tables should be read vertically rather than horizontally.

Table 14.9. Specified use "recreation", its related physical alterations and changes of hydromorphology. *Source:* Extracted from the case studies by the authors

Specified use	Physical alteration[a]	Changes of hydromorphology
Recreation (including angling, boat traffic)	*Rivers: Channels/beds* Shipping channels Creation of impounded lakes *Rivers: Banks/riparian zones* Marinas Tow path Land reclamation Buildings Landing stages/marinas Roads Picnic/recreation areas	*Rivers: Hydrological changes/impacts* No changes explicitly noted in the case studies *Rivers: Morphological changes/impacts* Change of bank structure Caving in of banks
		Other impacts Change of water quality

[a] The physical alterations and hydromorphological changes in the same row of the table do not necessarily or exclusively correspond to one another. The tables should be read vertically rather than horizontally.

Table 14.10. LAWA criteria for the identification of significant (physical) pressures, Germany. *Source:* Case study on the River Lahn, D

Specified uses	Significant	Not significant
Navigation	1. >10% impounded river length at mean low water flow 2. Navigation of passengers, goods-traffic, national waterway 3. Impassable artificial barriers with a height >30 cm 4. Proportion of river length with discharge acceleration with: – ratio profile depth to profile width ≥ 1:4; – bank (single or both sides) ≥ 10% total length with bank impairments; and – longitudinal profile ≥ 70% stretched or straightened	1. ≤10% impounded river length at mean low water flow 2. Recreational uses, motor boats, rowboats and canoes 3. Artificial barriers with a height ≤30 cm, passable artificial barriers with a height >30 cm respectively 4. Proportion of river length with discharge acceleration with: – ratio profile depth to profile width < 1:4; – bank (single or both sides) < 10% total length with bank impairments; – longitudinal profile < 70% stretched or straightened
Flood protection	1. Flood protection structures (dams, dikes) located within a strip up to 100 m at the potential floodplain along the river and directly connected with the river or located within a strip of 40% of the adjacent potential flooded riparian zones 2. >50% dyke construction works at the free-flowing river length	1. Flood-protection structures (dams, dikes) located within a strip up to 100 m at the potential floodplain along the river, located at least outside of 40% of the adjacent potential flooded riparian zones 2. ≤50% dyke construction works at the free-flowing river length
Hydropower generation	1. >10% impounded river length at mean low water flow 2. Impassable artificial barriers with a height >30 cm 3. Proportion of river length with discharge acceleration with: – ratio profile depth to profile width ≥ 1:4; – bank (single or both sides) ≥ 10% total length with bank impairments; and – longitudinal profile ≥ 70% stretched or straightened 4. Intermitted flow regulation with flow spills	1. ≤10% impounded river lengthat mean low water flow 2. Artificial barriers with a height ≤30 cm, passable artificial barriers with a height >30 cm respectively 3. Proportion of river length with discharge acceleration with: – ratio profile depth to profile width < 1:4; – bank (single or both sides) < 10% total length with bank impairments; – longitudinal profile < 70% stretched or straightened 4. Flow regulation without spills
Agriculture/forestry	1. Tillage and grassland > 50% of the river length 2. Special crop > 3–5% of the river length 3. Impassable artificial barriers with a height >30 cm 4. >50% of the entire river length in the rural landscape is impaired in the adjacent land zone	1. Tillage and grassland ≤ 50% of the river length 2. Special crops ≤ 3–5% of the river length 3. Artificial barriers with height ≤30 cm, passable artificial barriers with height >30 cm 4. ≤50% of the entire river length in the rural landscape is agriculture-like impaired in the adjacent land zone

Table 14.10. *Continued*

Specified uses	Significant	Not significant
Water supply	1. Drafts > 10% of mean low water flow 2. Fluctuated discharge ≥ 10% of mean water flow 3. No minimum discharge (according to respective land regulations) in rivers 4. Without recharge >0.1 mean low water flow per single installation and >0.5 mean low water flow total 5. With recharge >0.3 mean low water flow per single installation	1. Drafts ≤ 10% of mean low water flow 2. Fluctuation of the discharge < 10% of mean water flow 3. Minimum discharge (according to respective land regulations) in rivers 4. Without recharge ≤0.1 mean low water flow per single installation and = 0.5 mean low water flow total 5. With recharge ≤0.3 mean low water flow per single installation
Urbanisation	1. Urban areas > 10–15% of the river length 2. >50% of the entire river length are urban with bank fixation	1. Urban areas ≤ 10–15% of the river length 2. ≤50% of the entire river length are urban with bank fixation

Table 14.11. Indicators of the intensity of physical alterations in the Dender River (B). *Source:* Case study on the River Dender, B

Modifications	Intensity of modification					
	No modification 0	Very limited 1	Limited modification 2	Mediocre modification 3	Heavy modification 4	Very heavy modification 5
Annual sediment transport to river (for each VHA-zone)	No sediment delivery	<1 000 ton ha^{-1}	1 000–2 000 ton ha^{-1}	2 000–3 000 ton ha^{-1}	3 000–4 000 ton ha^{-1}	>4 000 ton ha^{-1}
Annual cumulative sediment transport to river	No sediment delivery	<5 000 ton ha^{-1}	5 000–10 000 ton ha^{-1}	10 000–15 000 ton ha^{-1}	15 000–20 000 ton ha^{-1}	>20 000 ton ha^{-1}
% hardened surface along the banks	0%	0–20%	20–40%	40–60%	60–80%	80–100%
% agricultural area along the banks	0%	0–20%	20–40%	40–60%	60–80%	80–100%
Canalisation	None	Gabarit to 300 ton	Gabarit to 600 ton	Gabarit to 1200 ton	Gabarit >1200 ton	–
Artificial discharge	Percentage of waste water of discharge Q_{90} is 0%	Percentage of waste water of discharge Q_{90} is <20%	Percentage of waste water of discharge Q_{90} is 20–40%	Percentage of waste water of discharge Q_{90} is 40–60%	Percentage of waste water of discharge Q_{90} is 60–80%	Percentage of waste water of discharge Q_{90} is 80–100%
Artificial discharge	Percentage of withdrawals of discharge Q_{90} is 0%	Percentage of withdrawals of discharge Q_{90} is <20%	Percentage of withdrawals of discharge Q_{90} is 20–40%	Percentage of withdrawals of discharge Q_{90} is 40–60%	Percentage of withdrawals of discharge Q_{90} is 60–80%	Percentage of withdrawals of discharge Q_{90} is 80–100%
Locks	None	One lock over the entire length of the river	One lock for four river sections	One lock for three river sections	One lock for two river sections	One lock for each river section
Dams/water mills	None	One dam over the entire length of the river	One dam for four river sections	One dam for three river sections	One dam for two river sections	One dam for each river section
Intensity navigation (commercial)	None	<10 boats yr^{-1}	<100 boats yr^{-1}	<1 000 boats yr^{-1}	<10 000 boats yr^{-1}	>10 000 boats yr^{-1}
Intensity navigation (recreational)	None	<10 boats yr^{-1}	<100 boats yr^{-1}	<1 000 boats yr^{-1}	<10 000 boats yr^{-1}	>10 000 boats yr^{-1}

Table 14.11. *Continued*

Modifications	Intensity of modification					
	No modification 0	**Very limited 1**	**Limited modification 2**	**Mediocre modification 3**	**Heavy modification 4**	**Very heavy modification 5**
Change in sinuosity (winding factor of the river) between 1850 and 1990	0	−0.2–0	−0.4–0.6	−0.6–0.8	−0.8–1.0	>−1.0
Dredging operations	None	On 0–20% of the total length of the river section	On 20–40% of the total length of the river section	On 40–60% of the total length of the river section	On 60–80% of the total length of the river section	On 80–100% of the total length of the river section

Table 14.12. Indicators for the classification of physical alterations on the basis of GIS data in the Dalälven River (S). *Source:* Case study on the River Dalälven, S (after Lövgren et al. 2000)

Indicator	Unit	Theoretical interval	Interval found in the River Dalälven (stream orders 1–3)
Degree of regulation	%/100	0–1	0–0.85
Density of dams	No/km^2	0–*x*	0–3.39
Proportion of stream transformed to log-driving channel	%/100	0–1	0–1
Density of roads	km/km^2	0–?	0–4.63
Proportion of arable land	km^2/km^2	0–1	0–0.70
Proportion of densely populated area	km^2/km^2	0–1	0–0.60
Proportion of clearcutted area	km^2/km^2	0–1	0–0.67

Table 14.13. Control variables according to types of environmental modifications in the Sarre River (F). *Source:* Case study on the River Sarre, F

Control variables (inputs)	Types of environmental modification					
	1 **Artificial modification of bed and banks**	2 **Diversion of all or part of the water course**	3 **Presence of continuous obstacles connecting the two banks of the water course**	4 **Release of retained volumes from locks**	5 **Surface runoff, surface sealing**	6 **Reduction in the surface area of the flood expansion field**
Stream discharge	Reorganisation (increase in bankfull discharge) Transverse obstacles Disruption of the regime	Reduction of minimum flows and mean flows, flood control	Disruption of natural flow rate diversity Disruption of the regime	Disruption of the regime (flood control, minimum flow compensation) Disruption of natural flow rate diversity	Radical reduction in roughness (stream channel and floodplain), increase of flood peaks (discharge from valley sides)	Transformation of hydraulic conditions (reduced flood coverage, concentration of flow)
Sediment discharge	Change in the nature of the riverbed and banks	Reduction in the power of the water course (particularly for flows >BD)	Disruption of sedimentary transit	Removal of fine sediment, disruption of sediment discharge regime	Removal of bed load Additional delivery of eroded material	Concentration of erosion-transport-deposition phenomena
Riparian vegetation	Clear cutting Allochtonous species Monospecificity	–	Clear cutting around constructions Allochtonous plantations	Bank erosion	Transformation of roughness	Bed erosion and bank destabilisation
Valley gradient	Uplift of the alluvial plain, bed incision	–	–	–	–	Embankment, levelling

Table 14.14. Case study results on current ecological status. *Source:* Extracted from the case studies by the authors

Country	Case study	Water bodies	Biological quality elements				Physicochemical	Hydro-morphological	Current ecological status
			Fish	Benthic invertebrate fauna	Macrophytes /phytobenthos	Phytoplankton			
A	Wienfluss R.	WB 1 (upper, unimpacted)	WB 1: good	WB 1: good					WB 1: good
		WB 2 (downstream, impacted)	WB 2: poor	WB 2: poor					WB 2: poor
	Bregenzerach R.	WB 1 (hydrologically unaffected)	WB 1: good	WB 1: high					WB 1: good
		WB 2 and 3 (residual stretches with minimum flow allocation)	WB 2 and 3: moderate to good	WB 2: good					WB 2 and 3: good
				WB 3: good to moderate					
		WB 4 (hydropeaking)	WB 4 poor	WB 4: tending to moderate					WB 4: poor (indicated especially through fish fauna)
	Danube R.	1 WB	Moderate	Moderate					Moderate
B	Dender R.	18 WBs (Mark, Dender, Bellebeek rivers)							Mark: some sections with moderate and some sections with good Dender: mostly moderate ecological status; only 2 sections of good status and 1 with poor Bellebeek: mostly moderate

Table 14.14. *Continued*

Country	Case study	Water bodies	Biological quality elements				Physicochemical	Hydro-morphological	Current ecological status
			Fish	Benthic inver-tebrate fauna	Macrophytes /phytobenthos	Phytoplankton			
D	Elbe R.	WB 1: upper	WB 1: <good	WB 1: moderate					No explicit overall outcome mentioned
		WB 2: middle	WB 2: no result	WB 2: no result					
		WB 3: lower	WB 3: no result	WB 3: no result					
	Seefelder Aach R.	1 WB							No explicit overall outcome mentioned
	Lahn R.	WB 1: upper Lahn							No explicit overall outcome mentioned
		WB 2: upper WB	WB 2: good	WB 2: moderate					
		WB 3: middle							
		WB 4: lower							
		WB 5: lower							
	Ruhr R.	Unit 1	Unit 1: no clear result	Unit 1: no clear result					Unit 1: good
		Unit 2	Unit 2: good	Unit 2: no clear result					Unit 2: good
		Unit 3 (WB 1)	Unit 3: good	Unit 3: no clear result					Unit 3 (WB 1 and 2): <good
		Unit 3 (WB 2)							
	Mulde R.	7 WBs	WB 1: bad	WB 1: good					WB 1: bad
			WB 2: bad	WB 2: moderate					WB 2: bad

Table 14.14. *Continued*

Country	Case study	Water bodies	Biological quality elements				Physicochemical	Hydro-morphological	Current ecological status
			Fish	Benthic inver-tebrate fauna	Macrophytes /phytobenthos	Phytoplankton			
			WB 3 poor	WB 3: no result					WB 3: poor
			WB 4: poor	WB 4: bad					WB 4: bad
			WB 5: moderate	WB 5: good					WB 5: moderate
			WB 6: moderate	WB 6: bad					WB 6: bad
			WB 7: moderate	WB 7: moderate					WB 7: moderate
	Dhünn R.	1 WB	(at least) good	(at least) good					good
E	Lozoya R.	WB Group 1 (natural stream)	WB Group 1: good	WB Group 1: high	WB Group 1: high		WB Group 1: high/good		WB Group 1: high/good
		WB Group 2 (reservoirs)	WB Group 2: bad	WB Group 2: bad	WB Group 2: bad		WB Group 2: moderate/good		WB Group 2: bad
		WB Group 3a (regulated river)	WB Group 3a: poor	WB Group 3a: moderate/bad	WB Group 3a: poor		WB Group 3a: good/high		WB Group 3a: poor
		WB Group 3b (final regulated river)	WB Group 3b: poor	WB Group 3b: good	WB Group 3b: moderate		WB Group 3b: good/high		WB Group 3b: moderate
SF	Kemijärvi L.	7 WBs in 2 groups	Moderate/good	Moderate/good	Moderate		General conditions: good/high	Hydrology: significant changes Morphology: moderate changes	Good or moderate
F	Authie R.		Moderate	Moderate	Moderate		Moderate		Moderate
	Sarre R.	21 WBs (28 WBs identified and reduced to 21 after screening)	Element used but no summary of results	Element used but no summary of results	Element used but no summary of results	Element used but no summary of results			3 WBs: good 18 WBs: <good

Table 14.14. *Continued*

Country	Case study	Water bodies	Biological quality elements				Physicochemical	Hydro-morphological	Current ecological status
			Fish	Benthic invertebrate fauna	Macrophytes /phytobenthos	Phytoplankton			
	Rhone R.	15 WBs in 3 groups: – heavily modified (6 Wbs) – created (artificial) (7 Wbs) – not affected by hydromorphological changes (2 WBs)	Overall: moderate For 2 WBs (Bregnier Cordon and Pierre Benite): fairly good	For 2 WBs (Bregnier Cordon and Pierre Benite): fairly good		For 2 WBs (Bregnier Cordon and Pierre Benite): fairly good	Fairly good	Upper Rhone (from Swiss border to downstream of Génissiat): bad Upper Rhone (from River Fier to River Guiers then from River Guiers to Sault-Brénaz): mediocre From Sault-Brénaz to River Ain junction: medium From River Ain junction to downstream of Péage-de-Roussillon: bad	Upstream course down to R. Saône junction: fairly good In R. Rhone: from R. Saône junction AND n BPR: through Lyons to restitution of Pierre-Bénite branch channel: fairly good to mediocre Pierre-Bénite Branch channel; Rhone, BPR, and branch channel from Pierre- Bénite restitution to downstream of Péage-de-Roussillon restitution: mediocre
GR	Nestos R.	WB 1 (Upstream of dams) WB 2 (Reservoirs) WB 3 (Downstream of dams) WB 4 (Delta)					WB 1: high WB 2: good WB 3: good WB 4: good		Not possible to define current ecological status (lack of data) – assumed to be 'moderate'

Table 14.14. *Continued*

Country	Case study	Waterbodies	Biological quality elements				Physicochemical	Hydro-morphological	Current ecological status
			Fish	Benthic invertebrate fauna	Macrophytes /phytobenthos	Phytoplankton			
NL	Haringvliet Est.	3 WBs	All 3 WBs: poor						All 3 WBs: poor
	Hagmolen-beek-Hege-beek R.	WB 1 (upstream Hegebeek)	WB 1: good to moderate	WB 1: good to moderate			Defined		WB 1: good to moderate
		WB 2 (Hagmolenbeek and downstream Hegebeek)	WB 2: moderate to good	WB 2: low					WB 2: moderate to good for fish and low for benthic invertebrate fauna
	Loosdrecht L.	1 WB							Low
	Veluwerand-meren	1 WB							Moderate to good
NO	Suldalslagen R.	1 WB	Poor	Moderate	Phytobenthos: moderate Macrophytes: poor		Good/high	Hydrological regime: poor River continuity: poor Morphological conditions: poor	Poor
	Beiarn R.	WB A (almost pristine)	WB A: high	WB A: high	WB A: high		WB A: high	WB A: high	WB A: high
		WB B and C (stretches with reduced water flow	WB B and C: good	WB B and C: moderate	WB B and C: good		WB B and C: high/good	WB B and C: high for river conti-nuity/moderate for hydrological regime and morphological conditions	WB B and C: moderate

Table 14.14. *Continued*

Country	Case study	Water bodies	Biological quality elements				Physicochemical	Hydro-morphological	Current ecological status
			Fish	Benthic invertebrate fauna	Macrophytes /phytobenthos	Phytoplankton			
		WB D (Stretch with partly recovered water flow and parasite infestation)	WB D: moderate	WB D: good	WB D: good		WB D: high/good	WB D: high for river continuity/good for hydrological regime and morphological conditions	WB D: moderate
S	Emån R.	WB 1 (lake)	WB 1: moderate					WB 1: moderate	No explicit overall outcome mentioned
		WB 2	WB 2: –					WB 2: moderate	
		WB 3 (reservoir)	WB 3: bad					WB 3: moderate	
		WB 4	WB 4: good					WB 4: moderate	
	Dalälven R.	WB 1 (reservoir + down-stream reach)	For all 5 WBs: <good						WB 1: bad
		WB 2 (regulated lakes)							WB 2: bad
		WB 3 (regulated lakes)							WB 3: bad
		WB 4 (dams and HP plants)							WB 4: bad
		WB 5 (dams and HP plants)							WB 5: bad
	Umealven R.	18 WBs						Defined	All 18 WBs: bad
	Baltic coast	3 WBs							WB 1: <good WB 2: <good WB 3: good

Table 14.14. *Continued*

Country	Case study	Water bodies	Biological quality elements				Physicochemical	Hydro-morphological	Current ecological status
			Fish	Benthic invertebrate fauna	Macrophytes /phytobenthos	Phytoplankton			
UK (EW)	Kennet R. (Thames)	8 WBs	Good	Good					(*Appears to be*) good
	Tame R.	3 WBs						WB 1: poor WB 2: bad	No explicit overall outcome mentioned
	Sankey Brook	Sub area Sankey Sub area Rainford	Sankey: poor Rainford: poor	Sankey: poor Rainford: poor			Sankey: poor Rainford: poor		Sankey: poor Rainford: poor
	Great Ouse R.	WB 1: Great Ouse WB 2-3-4: Wissey (upper, middle, lower)	WB 1: moderate	WB 1: moderate	WB 1: poor				WB 1: moderate (towards lower end) WB 2: moderate WB 3: moderate WB 4: poor
UK (S)	Forth Est.	3 WBs	WB A: good or moderate WB B: moderate	WB A: good WB B: moderate	WB A: – WB B: <moderate	WB A: high (phytoplankton), high–good (macroalgae) WB B: high (phytoplankton), good (macroalgae)	WB A: good WB B: good	WB A: high/good WB B: high to moderate	WB A: good WB B: <good

Table 14.14. *Continued*

Country	Case study	Water bodies	Biological quality elements				Physicochemical	Hydro-morphological	Current ecological status
			Fish	Benthic invertebrate fauna	Macrophytes /phytobenthos	Phytoplankton			
			WB C: moderate	WB C: moderate	WB C: <moderate	WB C: good (phytoplankton), good (macroalgae)	WB C: good–moderate	WB C: high to <moderate	WB C: <good
	Tummel R.	54 WBs							WBs directly influenced by HP (33 WBs): moderate WBs with additional alteration of water temperature (4 WBs): poor Some WBs affected by HP (17 WBs): good (steepness of the area, fish remediation measures)
	Dee R.	42 WBs							Most WBs between the farthest upstream and farthest downstream structure of HP (28 WBs): less than good Rest 14 WBs: good
UK (NI)	Lagan R.	5 WBs							All 5 WBs: <good

Table 14.15. Case study results on failure to achieve good ecological status and HMWB provisional identification. *Source:* Extracted from the case studies by the authors

Country	Case study	Water bodies	Current ecological status	Reasons for failing to achieve Good Ecological Status	WBs identified as *provisionally HMWB*[a]
A	Wienfluss R.	WB 1 (upper, unimpacted)	WB 1: good		WB 2 (downstream impacted section)
		WB 2 (downstream, impacted)	WB 2: poor	Lack of habitat heterogenity, bank reinforcement	
	Bregenzerach R.	WB 1 (hydrologically unaffected)	WB 1: good	Hydropower peaking operations	
		WB 2 and 3 (Residual stretches after minimum flow allocation)	WB 2 and 3: good		
		WB 4 (hydropeaking)	WB 4: poor (indicated especially through fish fauna)		WB 4
	Danube R.	1 WB	Moderate	Dams, disruption of river continuum	1 WB
B	Dender R.	18 WBs (Mark, Dender, Bellebeek rivers)	Mark: some sections with moderate and some sections with good Dender: mostly moderate ecological status; only 2 sections of good status and 1 with poor Bellebeek: mostly moderate	Weirs, straightening, channel maintenance, embankments	3 WBs of the Dender (WBs 6-7-8) Entire Bellebeek and Mark are natural
D	Elbe R.	WB 1: upper		Bank impairments, detachment of oxbow lakes	WB 1
		WB 2: middle			No information on other WBs
		WB 3: lower			
	Seefelder Aach R.	1 WB		Agriculture	1 WB
	Lahn R.	WB 1: upper Lahn		Barriers to migration, intermittent flow	
		WB 2: upper			WB 2
		WB 3: middle			No information on other WBs

Table 14.15. *Continued*

Country	Case study	Water bodies	Current ecological status	Reasons for failing to achieve Good Ecological Status	WBs identified as *provisionally HMWB*[a]
		WB 4: lower			
		WB 5: lower			
	Ruhr R.	Unit 1	Unit 1: good	Impounded lakes, weirs/river continuum disruption, substrate siltation, critical oxygen depletion, high ammonia	Unit 3 (WB 1 and 2)
		Unit 2	Unit 2: good		
		Unit 3 (WB 1 and 2)	Unit 3 (WB 1 and 2): <good		
	Mulde R.	7 WBs		Mainly physicochemical impacts	3 WBs (WB 2 – urban area, WB 3 – transition zone of central lowland, WB 5 – urban area)
	Dhünn R.	1 WB	Good		None
E	Lozoya R.	WB Group 1 (natural streams): 4 WBs	WB Group 1: high/good	Reservoirs, river regulation	Not mentioned
		WB Group 2(reservoirs): 5 Ws	WB Group 2: bad		(WB Ggroup 2: 5 WBs)
		WB Group 3 (regulated river): 1 W	WB Group 3a: poor		(WB Group 3a: 1 WB)
		WB Group 3b (final regulated river): 2 WBs	WB Group 3b: moderate		(WB Group 3b: 2 WBs)
SF	Kemijärvi L.	7 WBs in 2 groups	Good or moderate	Lake regulation for hydropower	Lake Kemijärvi (not clear whether this is only group 1 or both groups)
F	Authie R.	–	Moderate		Not mentioned
	Sarre R.	21 WBs (28 WBs identified and reduced to 21 after screening)	3 WBs: good 18 WBs: <good	Fish dysfunction, invertebrate degradation due to mill dams, minimum flow deficit, canalisation	18 WBs

Table 14.15. *Continued*

Country	Case study	Water bodies	Current ecological status	Reasons for failing to achieve Good Ecological Status	WBs identified as *provisionally HMWB*[a]
	Rhone R.	15 WBs in 3 groups: – heavily modified (6 WBs) – created (artificial) (7 WBs) – not affected by hydro-morphological changes (2 WBs)			Not mentioned (6 WBs)
GR	Nestos R.	WB 1 (Upstream of dams)	Not possible to define current ecological status due to lack of data – assumed to be 'moderate'	Water level regulation in reservoirs and hydropower dam operation	Not mentioned (WB 1, WB 3 and WB 4) (WB 2 reservoirs as artificial)
		WB 2 (Reservoirs)			
		WB 3 (Downstream of dams)			
		WB 4 (Delta)			
NL	Haringvliet Est.	3 WBs	All 3 WBs poor	Lack of salinity gradient due to sluice, high contaminant level	All 3 WBs
	Hagmolenb. - Hegebeek R.	WB 1 (upstream Hegebeek)	WB 1: good to moderate	Land drainage, catchment disconnection, altered bank profile	
		WB 2 (Hagmolenbeek and downstream Hegebeek)	WB 2: moderate to good for fish and low for benthic invertebrate fauna		WB 2
	Loosdrecht L.	1 WB	Low	Water level management and unnatural condition of banks	1 WB
	Veluwerand-meren	1 WB	Moderate to good	Water level management and unnatural condition of banks	1 WB
NO	Suldalslagen R.	1 WB	Poor	Change in flow regime, reduced flood frequency, no ice jam events, changed substrate, increased moss	1 WB

Table 14.15. *Continued*

Country	Case study	Water bodies	Current ecological status	Reasons for failing to achieve Good Ecological Status	WBs identified as *provisionally HMWB*[a]
	Beiarn R.	WB A (almost pristine)	WB A: high	Reduced water flow, parasite infestation (for WB D)	Not mentioned (WB B, C and D)
		WB B and C (Stretches with reduced water flow)	WB B and C: moderate		
		WB D (Stretch with partly recovered water flow and parasite infestation)	WB D: moderate		
S	Emån R.	WB 1 (lake) WB 2:			
		WB 3 (reservoir)		WB 3 identified as provisionally HMWB due to 2 HP plants. It will not achieve good status due to: – very large height of fall of HP plants; – high costs of building fish ladders; and, – destruction of suitable fish habitats	WB 3
		WB 4			
	Dalälven R.	WB 1 (reservoir+downstream reach)	WB 1: bad	Disrupted river continuum, high regulation amplitude	All 5 WBs
		WB 2 (regulated lakes)	WB 2: bad		
		WB 3 (regulated lakes)	WB 3: bad		
		WB 4 (dams and HP plants)	WB 4: bad		
		WB 5 (dams and HP plants)	WB 5: bad		
	Umealven R.	18 WBs	All 18 WBs: bad	Hydropower dams, run-off-river impoundment, change in flow regime	All 18 WBs
	Baltic coast	3 WBs	WB 1: <good		Not mentioned (WB 1 and 2)
			WB 2: <good		
			WB 3: good		

Table 14.15. *Continued*

Country	Case study	Water bodies	Current ecological status	Reasons for failing to achieve Good Ecological Status	WBs identified as *provisionally HMWB*[a]
UK (EW)	Kennet R. (Thames)	8 WBs	(*Appears to be*) good (based on available reference conditions; there are no 'natural' large chalk streams to consider as reference conditions)	Weirs as barriers to connectivity	All 8 WBs (even though biology shows good status but this could be biased since biological samples are taken at sites of good habitat)
	Tame R.	3 WBs		Pollution, altered flow regime and physical habitat alteration	3 WBs as borderline HMWB (but only WB 1 and WB 2 have proceeded to the designation assessments)
	Sankey Brook	3 groups (according to use): – urban (6 Wbs) – semi-urban (4 Wbs) – rural (7 WBs)	Sub area Sankey: poor Sub area Rainford: poor	Poor water quality, low habitat diversity (bank and in-channel engineering, lack of meanders, culverts and weirs, lack of floodplain)	Not mentioned (all 17 WBs?)
	Great Ouse R.	WB 1: Great Ouse	WB 1: moderate (towards lower end)	Impoundment, lack of connectivity, bank diversity, fast flowing areas	All WBs
		WB 2–4: Wilsey (upper, middle, lower)	WB 2 moderate WB 3: moderate WB 4 poor		
UK (Scot)	Forth Est.	WB A: lowstream	WB A: good	Dredging, land reclamation, port facilities	WB B and C
		WB B: middlestream	WB B: <good		
		WB C: upstream	WB C: <good		

Table 14.15. *Continued*

Country	Case study	Water bodies	Current ecological status	Reasons for failing to achieve Good Ecological Status	WBs identified as *provisionally HMWB*[a]
	Tummel R.	54 WBs	WBs directly influenced by HP (33 WBs): moderate WBs with additional alteration of water temperature (4 WBs): poor Some WBs affected by HP (17 WBs): good (steepness of the area, fish remediation measures)	River continuum disruption, change in river profile, mechanical damage of fauna	37 WBs: WBs directly influenced by HP (33 WBs) and WBs with additional alteration of water temperature (4 WBs)
	Dee R.	42 WBs	Most WBs between the farthest upstream and farthest downstream structure of HP (28 WBs): less than good Rest 14 WBs: good	Hydrological changes due to HP, sediment starvation and changes in vegetation downstream of catchwaters	28 WBs
UK (NI)	Lagan R.	5 WBs	All 5 WBs: <good	Although all 5 WBs fail good ES, it is likely that WB 1 and 2 failed due to pollution pressures rather than physical modification WB 3, 4 and 5 have moderate/poor status as a result of significant physical modification	3 WBs (WB 3, 4, 5) to be further considered (therefore provisionally as HMWB)

[a] When no results were mentioned in the case studies, the authors of the synthesis made assumptions on the provisional identification of HMWB based on the results of the current ecological status classification (assumed results are mentioned in brackets).

Table 14.16. Hydromorphological elements for defining ecological status in the Hagmolenebeek River (NL). *Source:* Case study on the River Hagmolenbeek-Hegebeek, NL

	Hydrology		River continuity (No. of weirs)	Morphology longitudinal	Morphology lateral		Structure and substrate streambed
	Quantity of flow	Dynamics of flow		River depth and width variation	River depth and width variation	Structure of the riparian zone	
Hegebeek upstream 20-11-4 (2 km)	From incidentally/periodically stagnant water to dry for short periods	Variation in flow is available: almost permanent; two to three different 'flow patterns' are present	Yes (0)	Present: natural meandering pattern	Present: natural meandering pattern	Natural vegetation: stream is almost completely shady	Silt, sand, gravel, organic matter and dead wood
Hegebeek downstream 20-5-1 (3 km)	From incidentally/periodically stagnant water to dry for short periods	Variation in flow is absent: almost permanent; only one 'flow pattern' present	No (6)	Absent due to canalisation	Absent due to normalisation	Semi-natural vegetation on one side of the stream: half of stream is shady	Silt, sand, organic matter
Hagmolenbeek 20-5 (22 km)	From incidentally/periodically stagnant water to dry for short periods	Variation in flow is absent: almost permanent only one 'flow pattern' present	No (14)	Absent due to canalisation	Absent due to normalisation	Semi-natural vegetation on one side of the stream: half of stream is shady	Silt, sand, organic matter

Table 14.17. Classification system for rivers from the case study on the Emån River (S). *Source:* Case study on the River Emån, S

Element		High status	Good status	Moderate status	Poor status	Bad status
Fish fauna[a]		Optimal or near optimal population of fishes living in streaming water considering the natural conditions	Occurrence and recruitment of fishes living in streaming water is fairly good, but not optimal because of some kind of negative influence. The population might be recovering after a former influence	Occurrence and recruitment of fishes living in streaming water is fairly good, but not optimal because of some kind of negative influence. It is uncertain if the population is recovering after an influence or degenerating because of an influence	Negative influence on the population. Unnatural composition of species. Further disturbances might exterminate the population	Strong negative influence on the population. Further disturbances might exterminate the population
Macroinvertebrates[b]	Number of taxa	>42	35–42	25–34	15–24	<15
	Number of ind. m^{-2}	>3 800	2 000–3 800	510–2 000	100–500	<100
	Shannon diversity index	>3.71	2.97–3.71	2.22–2.97	1.48–2.22	<1.48
Permanent modifications[c] (straightening, channelisation, broadening or dredging of the river channel or filling of a part of a river channel)		0–10%	10–25%	25–50%	50–75%	>75%
Modification of the flow[c] (anthropogenic modification, which influences the discharge of the river)		None or some surface run off water	Reservoirs, no biological negative effect	Regulation, <25% difference between high and low discharge during the day/week	Regulation (short time), >25% difference between high and low discharge during day/week	Periodically dry river channels
Fragmentation[c] (fragmentation caused by artificial obstacles)		0	>0–25%	25–50%	50–75%	>75%
Land use[c] (impacts such as clear-cut areas, arable land and settlements)		<10%	>10–20%	>20–40%	>40–60%	>60%

[a] These criteria are from the method used by the County Administrative Board in Jönköping when assessing ecological quality of fish in rivers (Tärnåsen and Haag 2000).
[b] These criteria are from the method used by the Emån Water Association by the assessment of ecological quality of macroinvertebrates (Pröjts 2000). The criteria have been modified to some extent in order to agree with the WFD.
[c] These criteria are from System Aqua, which is a method used by the Swedish Environmental Protection Agency to assess ecological quality (Reinholdson 2001). The criteria have been modified to some extent in order to agree with the WFD.

Table 14.18. Classification of ecological status in the Umealven River (S). *Source:* Case study on the River Umealven, S

Type of criteria	Ecological status				
	Bad	Poor	Moderate	Good	High
Ecosystem criteria					
Reduction in the width of the riparian zone	–	<50% reduction	<25% reduction	None or little	None or very little
Presence of plant zonation	–	–	Present	Present	Present
Community criteria					
Species richness of riparian plants	–	<30% reduction	<20% reduction	<10% reduction	Natural or near natural
Reduction in riparian plant cover	–	<50% reduction	<40% reduction	<25% reduction	Natural or near natural
Species richness of aquatic invertebrates	–	–	–	–	–
Artificial barriers to upstream migration	–	–	None, or functioning passes present	None, or functioning passes present	None
Artificial barriers to downstream migration	–	–	Migration possible during critical periods	Only minor barriers	None
Species and population criteria					
Populations of anadromous fish species	–	–	Present	Present and viable	Present and viable
Hydromorphological criteria					
Seasonal changes in flow	–	Some kind of spring flood, followed by lower flows	Spring flood >50% of natural in height and duration, autumn and winter flows increased by <50%	Spring flood >90% of natural in height and duration, autumn and winter flows increased by <10%	Natural
Diversions of flow	–	–	>75% of natural	None or small	None

Table 14.19. Threshold values for biological elements and ecological quality ratios in Lake Kemijärvi (SF). *Source:* Case study on the Lake Kemijärvi, SF

Biological quality element	Ecological status			Reference lake	Lake Kemijärvi			
	Moderate	Good	High	Value	Nat. value	Pres. value	*EQR*	Ranking
Macrophytes								
Number of species (*n*)	30	40	50		46	37	0.80	High
Area of flooded *Carex* vegetation (ha)	–	–	–		179	14	0.08	Moderate
Vertical extension of *Carex* zone (m)	0.4	0.6	0.8		0.85 (3.04)[a]	0.45 (1.01)[a]	0.30–0.53	Moderate
Presence of frost sensitive isoetids (*n*)	1	2	3		2	1	0.50	Moderate
Macrozoobenthos								
Number of typical taxons (*n*)	>15	>22	>29	35.9	–	35	0.97	High
Number of typical taxonomical groups (*n*)	>9	>13	>17	20.6	–	19	0.92	High
Number of taxons (*n*)	>19	>29	>38	47.14	–	42	0.89	High
Presence of frost sensitive species (*n*)	1	2	3		3	2	0.67	Good/ moderate
Fish fauna								
Number of species (*n*)	14	16	18	18	15–22	18	1.00	High
Presence of migratory species	No	Yes[b]	Yes[c]		Yes	No[d]	–	Moderate

Reference lake represents average values of littoral species from ten Finnish lakes; *EQR* = Ecological Quality Ratio.

[a] Maximum values in parenthesis.

[b] Migration possible to tributary.

[c] Migration possible to sea.

[d] Anadromous species.

Table 14.20. Classification of ecological status in the Dalälven River (S). *Source:* Case study on the River Dalälven, S

Type of criteria		Target unit	Ecological status				
			Bad	Poor	Moderate	Good	High
1	River continuum	*River catchment*	Negligible passage	Certain passage	Moderate passage	Well functioning passage	Natural passage
2	Degree of fragmentation	*Main river or catchment >100 km²*	>75%	<75%	<50%	<25%	None
3	Degree of exploitation of hydropower potential (height or effect)	*Catchment*	>95%	<95%	<75%	<50%	None
4	Barrier effect						
	% of potential reach with reproduction areas	*Catchment*	<10%	<25%	<75%	>75%	Natural
5	Effects on flow regime						
	Natural annual variation remaining (spring and autumn floods)	*River*	<25%	>25%	>50%	>75%	Natural
	Water discharge (% of natural)	*River*	<5%	>5%	>15%	>30%	Close to natural
	Proportion of natural stream reaches (%)	*River*	<25%	>25%	>50%	>75%	Close to natural
6	Type of regulation						
	Dry reach, permanent	*Downstream dam*	Bad	–	–	–	–
	Dry reach, minimum discharge (% natural)	*Downstream dam*	*<5%	*>5%	*>15%	*>30%	Close to natural
	Short-term fluctuations, weekly	*Run-off-river impoundment*	–	–	–	At most (see *)	None
	Short-term fluctuations, daily	*Run-off-river impoundment*	–	–	At most (see *)	–	None
	Short-term fluctuations, hourly	*Run-off-river impoundment*	–	At most (see *)	–	–	None
7	Regulation amplitude	*Reservoirs*	>4 m	<4 m	<2 m	<1 m	Natural
		Regulated lakes	>2 m	<2 m	<1 m	<0.5 m	Natural
8	Possibilities for fish migration						
	Turbine type	*Hydropower dam*	Francis	Francis	Francis	Kaplan	None
	Mechanical leading of fish away from turbine intake	*Hydropower dam*	Missing	Poorly functioning	Moderately functioning	Well functioning	Natural
	Bypass for fish	*Hydropower dam*	Missing	Poorly functioning	Moderately functioning	Well functioning	Natural
	Fish ladders or bypass	*Dam*	Missing	Poorly functioning	Moderately functioning (>50%)	Moderately functioning (>75%)	Natural

Table 14.20. *Continued*

Type of criteria	Target unit	Ecological status				
		Bad	Poor	Moderate	Good	High
9 Canalisation						
Stream bed cleared of stream thresholds, rocks, large stones etc. (former logging passages)	*Reach longer than 1 km*	–	–	Obviously cleared	Not obviously cleared	Natural
Channelled with artificial beach/quay	*Reach longer than 1 km*	–	–	Artificial beach/quay	Nature-like beach	Natural

Table 14.21. Case study results on the HMWB provisional identification and designation process. *Source:* Extracted from the case studies by the authors

Country	Name of case study	Number of WBs identified as provisional HMWB	Number of WBs designated as HMWB	Main reasons for designation (a) refers to designation test Art. 4 (3)(a) of the WFD (b) refers to designation test Art. 4 (3)(b) of the WFD
A	Wienfluss R.	WB	1 WB	Not mentioned
	Bregenzerach R.	WB	Not designated as HMWB	GES can be achieved without significant adverse effects (tunnel for hydropeaking flow). In other situations, or under different socio-economic conditions, case specific analyses are necessary and eventually may result in the designation of hydropeaking sections as heavily modified
	Danube R.	WB	1 WB	(a) significant adverse effects on use (urbanisation, no alternatives) + costly measures (b) no better env. option (hydropower)
B	Dender R.	3 WBs	3 WBs	(a) significant adverse effects on uses (adverse effects on navigation are most important); proposed erosion control measures do not adversely affect agriculture but cannot lead to GES on their own (b) no better env. option (navigation); technically not feasible (flood protection); alternatives for agriculture are possible but not sufficient on their own to achieve GES (potential social impacts on farmers); alternatives for recreation are possible but are opposite to the line of policy
D	Elbe R.	1 WB	Not designated as HMWB	(a) no significant adverse effects on use (navigation) (b) better environmental option (train instead of navigation) costs are not disproportionate, low profitability of inland navigation
	Seefelder Aach R.	1 WB	Not designated as HMWB	(a) no significant adverse effect on use (small % of agricultural land affected in the catchment) (b) no disproportionate costs
	Lahn R.	1 WB	Not designated as HMWB	(a) no significant adverse effects on use (non-profitable hydropower stations, inactive weirs)
	Ruhr R.	1 Unit (core region: 2 WBs)	1 WB (part of it) + stretch downstream of core region (harbour)[a]	(a) significant adverse effects on uses (b) disproportionate costs
	Mulde R.	3 WBs	2 WBs[b]	(a) significant adverse effect on use (urbanisation) (b) no alternatives to existing use (urbanisation)
	Dhünn R.	None	None	

Table 14.21. *Continued*

Country	Name of case study	Number of WBs identified as provisional HMWB	Number of WBs designated as HMWB	Main reasons for designation (a) refers to designation test Art. 4 (3)(a) of the WFD (b) refers to designation test Art. 4 (3)(b) of the WFD
E	Lozoya R.	Not mentioned		(a) significant adverse effect on use (water supply) (b) no alternatives to existing use (water supply)
		(Group 2 (reservoirs): 5 WBs)	Group 2 (reservoirs): 5 WBs	
		Group 3a and b (regulated river): 3 WBs	Group 3a and b (regulated river): 3 WBs	
SF	Kemijärvi L.	1 WB	1 WB	(a) significant adverse effect on use (hydropower, flood protection) (b) no better environmental option (hydropower), disproportionately costly (flood protection)
F	Authie R.	Not mentioned	Whole river[c]	(a) significant adverse effect on use (poplar plantation provides income to owners); measures would not lead to GES
	Sarre R.	18 WBs	5 WBs[d]	(a) significant adverse effect on use (urbanisation, canalisation) Grading value of parameters after rehabilitation was calculated for all provisionally identified HMWBs. (WB is designated HMWB, if parameter value after rehabilitation is less than 75% of the maximum ecological status.)
	Rhone R.	Not mentioned (6 WBs)	5 WBs	(a) proposed measures are not enough to restore fully active hydromorphological performance
GR	Nestos R.	Not mentioned (3 WBs + 1 WB-reservoir as artificial)	2 WBs as heavily modified + 1 WB (reservoir) as artificial	(a) significant adverse effects on use (hydropower, agriculture) (b) disproportionate costs for alternatives, flooding agricultural areas is not socially acceptable
NL	Haringvliet Est.	3 WBs	3 WBs	(b) technical feasibility of alternative options questionable, disproportionate costs
	Hagmolenbeek-Hegebeek R.	1 WB	1 WB	(a) significant adverse effects on use (agriculture, no alternative) (b) disproportionate costs GES cannot be achieved without restoration of hydrology

Table 14.21. *Continued*

Country	Name of case study	Number of WBs identified as provisional HMWB	Number of WBs designated as HMWB	Main reasons for designation (a) refers to designation test Art. 4 (3)(a) of the WFD (b) refers to designation test Art. 4 (3)(b) of the WFD
	Loosdrecht L.	1 WB	1 WB	(a) significant adverse effects on the use/wider environment (habitation, water supply) (b) no better environmental option
	Veluwerandmeren	1 WB	1 WB	(a) significant adverse effects on the use/wider environment (habitation) (b) disproportionate costs
NO	Suldalslagen R.	1 WB	1 WB final designation not conclusive[e]	(a) significant adverse effect on use/wider environment (hydropower, floodplains) (b) disproportionate costs
	Beiarn R.	Not mentioned (2 WBs)	2 WBs	(a) significant adverse effect on use (hydropower) (b) no better environmental option (hydropower), disproportionate costs
S	Emån R.	1 WB	1 WB	(a) significant adverse effect on use (hydropower) (b) no better environmental option (hydropower)
	Dalälven R.	All 5 WBs	5 WBs	(a) significant adverse effect on use (hydropower) (b) no better environmental options
	Umealven R.	18 WBs	18 WBs	(a) significant adverse effect on use (hydropower) (b) no better environmental options It is impossible to achieve GES in 15 years (river classified as bad ecological status)
	Baltic coast	Not mentioned (2 WBs-study areas)	2 WBs-study areas	(a) significant adverse effects on use (navigation, recreation) (b) no real alternatives
UK (EW)	Kennet R. (Thames)	All 8 WBs	1 WB (stretch from Hungeford to Newbury)	(a) significant adverse effects on use (b) costs outweigh benefits slightly
	Tame R.	3 WBs as borderline HMWB (but only WB 1 and WB 2 have preceeded to the designation assessments)	2 Wbs: WB 1 WB 2 (part of it)	(a) and (b) necessary restoration measure technically not feasible, disproportionate costs, no real alternatives

Table 14.21. *Continued*

Country	Name of case study	Number of WBs identified as provisional HMWB	Number of WBs designated as HMWB	Main reasons for designation (a) refers to designation test Art. 4 (3)(a) of the WFD (b) refers to designation test Art. 4 (3)(b) of the WFD
	Sankey Brook	Not mentioned	Not designated as HMWB	Water quality problems more important than physical modifications
	Great Ouse R.	1 WB	Not designated as HMWB	(a) and (b) Adding non navigable channels as 'natural habitats' has no significant adverse effect on use, is technically feasible and not disproportionate costly. However, it is uncertain if full GES can be achieved
	(A) Main Ouse			
	(B) Wissey	3 WBs	3 WBs	(a) significant adverse effect on use (flood protection, habitation) (b) no alternative option
UK (S)	Forth Est.	2 WBs	1 WB	(a) significant adverse effect on use (harbour) (b) no better env. options (navigation), disproportionately high costs (agriculture) 1 WB not designated as HMWB since reclamation of agricultural production is not disproportionately costly
	Tummel R.	37 WBs	37 WBs	(a) significant adverse effect on use/wider environment (hydropower, wetlands) (b) no better environmental option (hydropower), disproportionate costs
	Dee R.	28 WBs	21 WBs (in 4 WB groups)	(a) significant adverse effect on use/wider environment (hydropower, wetlands) (b) no better environmental options, disproportionate costs
UK (NI)	Lagan R. (incomplete)	3 WBs	2 WBs 1 WB designation not conclusive	(a) significant adverse effect on uses (b) no alternatives

[a] 'Clear-cut situation" quantitative analysis not necessary due to the importance of the inland harbour.
[b] 1 WB not designated because it was affected mainly by water quality problems.
[c] Designation due to agricultural impacts (diffuse pollution and erosion).
[d] 5 WBs designated, if proposed plan can be funded. In case of no increase in budget, there would be more WBs designated.
[e] Restored 'naturalised flow' could be realistically implemented, but it is uncertain whether this would ensure salmon population recovery.

Table 14.22. Case study results on Article 4(3)(a) – Assessment of restoration measures. *Source:* Extracted from the case studies by the authors

Country	Name of case study	Restoration measures to achieve GES	Positive effects of restoration measures	Costs of (setting up) restoration measures (not economic impact of measures)	Significant adverse effects on specified use? *Y(es)/N(o), Methods used*	Significant adverse effects on wider environment? *Y(es)/N(o), Methods used*	Remarks
A	Wienfluss R.	No information					
	Bregenzerach R.	Bypassing the surge through a tunnel directly into the Lake Constance	Measure will lead to Good Ecological Status		N	N	Restoration measures already in place; case study also contains details on the effectiveness of the measures on the ecological status
	Danube R.	1. Reestablishment of free flowing river sections		Considerable amount of money is necessary for implementing the defined measures; questionable if current users are able to contribute to the restoration	1. Significant reduction of hydropower generation (50% loss)		
		2. Reestablishment of dynamic floodplains along the newly created free flowing section			2. Transfer from forest and agricultural land to floodplains of lower economic value		
		3. Reestablishment of longitudinal and lateral connectivity			3. Increased maintenance work for navigation channel		

Table 14.22. *Continued*

Country	Name of case study	Restoration measures to achieve GES	Positive effects of restoration measures	Costs of (setting up) restoration measures (not economic impact of measures)	Significant adverse effects on specified use? *Y(es)/N(o), Methods used*	Significant adverse effects on wider environment? *Y(es)/N(o), Methods used*	Remarks
B	Dender R.	1. Removing locks			1. *Navigation*: Y, water level will be too low to make navigation (commercial and recreational) possible *Flood protection*: Y, increase of the risk of flood for surrounding areas		
		2. Rewinding of river			2. *Navigation*: Y *Urbanisation, agriculture*: Y, take up of space	2. Protected areas downstream seriously affected by change of river course	
		3. Removal of embankments			3. *Flood protection*: Y, rise of flood risk		
		4. Stop water abstractions			4. *Navigation*: Y, in the summer, water level will be too low to enable navigation *Urbanisation*: Y, factories will have to change their production process to obtain zero discharge		
		5. Stop dredging			5. *Navigation*: Y, due to sedimentation *Flood protection*: Y, due to storage capacity decrease		

Table 14.22. *Continued*

Country	Name of case study	Restoration measures to achieve GES	Positive effects of restoration measures	Costs of (setting up) restoration measures (not economic impact of measures)	Significant adverse effects on specified use? *Y(es)/N(o), Methods used*	Significant adverse effects on wider environment? *Y(es)/N(o), Methods used*	Remarks
		6. Erosion control measures			6. N, no significant adverse effect on agriculture		
D	Elbe R.	1. *Scenario A* (Maintenance of navigation) 2. *Scenario B* (Modification of the pressure): *only this scenario* was taken into consideration since the abandonment of navigation is not necessary to achieve GES	2. Decrease of 10–30 cm of navigable depths Development of natural hydromorphology in the long term		2. N *Business level* For 30–50% of the year, a mean (not a maximum) use of capacity of ships is not possible → at these times income losses of 5-22% are expected *Regional level* The share of goods transported by cargo ships in the area of the Elbe is only approx. 1%, clearly below the German average of 12%. Inland navigation on the Elbe is therefore of secondary importance compared to the demand at a regional level → no significant adverse effect on navigation	2. Y A removal of riverbed reinforcements could lead to an erosion due to increasing flow velocity. Impacts on the ecosystem (particularly the floodplain) are possible as a consequence of an induced lowering of the groundwater level	

Table 14.22. *Continued*

Country	Name of case study	Restoration measures to achieve GES	Positive effects of restoration measures	Costs of (setting up) restoration measures (not economic impact of measures)	Significant adverse effects on specified use? *Y(es)/N(o), Methods used*	Significant adverse effects on wider environment? *Y(es)/N(o), Methods used*	Remarks
		3. *Scenario C* (Abandonment of navigation of passengers and goods traffic)					
	Seefelder Aach R.	1. *Scenario A* (Maintenance)			2.–3. *Scenario B and C*: Loss of agricultural land on the river bank and in the whole floodplain The loss of agricultural land due to restoration measures will be 0.6% compared to the whole catchment area of the Seefelder Aach for Scenario B and about 1.6% for Scenario C. Regarding the long-term development of the amount of agricultural area, and according to the long-term view of the WFD, it can be assumed that the affected land can be compensated by the availability of land at another place A loss of agricultural land up to 10% is not a significant impact	2.–3. *Scenario B and C* Adverse effects for the wider environment are not expected	For the question of significance, one can put the focus on a local or on a regional level. An assessment on a local level focuses on the economics of the business of a single farm and requires detailed information such as the percentage of affected land in relation to the total farm size, farm structures, etc. Considering the difficulty in collecting such information, focus on a local level within the designation process in the case of agriculture is not recommended. Hence, the significance of impacts was seen on a regional level

Table 14.22. *Continued*

Country	Name of case study	Restoration measures to achieve GES	Positive effects of restoration measures	Costs of (setting up) restoration measures (not economic impact of measures)	Significant adverse effects on specified use? *Y(es)/N(o), Methods used*	Significant adverse effects on wider environment? *Y(es)/N(o), Methods used*	Remarks
		2. *Scenario B* (Modification): – creation of riparian buffer strips – removal of bank and channel bottom reinforcements – optimisation of management practice – conversion of arable land into pasture – lower intensity of land use – re-connection of oxbows 3. *Scenario C* (Abandonment): – abandonment of agricultural land use in the floodplain area – removal of bank and channel bottom reinforcements – re-connection of oxbows					
	Lahn R.	1. *Scenario A* (Maintenance of hydropower)	1 Scenario A was not evaluated because it would not lead to GES			No adverse effects	

Table 14.22. *Continued*

Country	Name of case study	Restoration measures to achieve GES	Positive effects of restoration measures	Costs of (setting up) restoration measures (not economic impact of measures)	Significant adverse effects on specified use? *Y(es)/N(o), Methods used*	Significant adverse effects on wider environment? *Y(es)/N(o), Methods used*	Remarks
		2. *Scenario B* (Modification of hydropower): – removal of weirs no longer used – building fish migration passes at weirs *without* use of hydroelectric power – building fish migration passes at weirs *with* the use of hydroelectric power – modification of the use (requirements to minimum rate of flow)			2. *Local level* At the business level, production costs are higher than reimbursement. Actual use seems not to be power generation Investment costs with and without ecological requirements are calculated (effect of ecological requirements on production costs) *National level* Considering the % of power generation of small HP plants in relation to total electricity consumption in D, loss of power is not significant → Effects on use are partially significant		
		3. *Scenario C* (Abandonment hydropower)			3. Abandonment of HP is a significant effect (clear-cut case)		

Table 14.22. *Continued*

Country	Name of case study	Restoration measures to achieve GES	Positive effects of restoration measures	Costs of (setting up) restoration measures (not economic impact of measures)	Significant adverse effects on specified use? *Y(es)/N(o), Methods used*	Significant adverse effects on wider environment? *Y(es)/N(o), Methods used*	Remarks
	Ruhr R.	1. *Scenario A* (close to natural status): – abandon impounded lakes – replace weirs with artificial riffles – remove bank fixations locally – establish alluvial forests – reattach oxbow lakes	1. Achieve dynamic streambed migration Develop spawning grounds and habitats		1. *Scenario A* – Total loss of hydropower production – Expected abandonment of passenger navigation in WB 2 (loss of landscape attractivity, missing rentability) – Significant restriction or complete abandonment of maintenance navigation – Significant loss of agricultural area	1. Abandonment of impounded lakes will decrease groundwater level and can cause oxbow lakes to dry in case of sludge removal. Material will be probably deposed in the floodplain	*To define restoration measures*: 1. The impact and resulting effects (physical and biological) were listed for each WB 2. Sub-ordinate targets were developed (reduce eutrophication, re-establish river continuum, restore natural stream bed) 3. Different scenarios (solution alternatives) of measures were developed. Scenarios consider only realistic factors (do not challenge existing uses) *To assess adverse effects on uses*: 1. Criteria and levels of significance for adverse effects are defined 2. Qualitative description of adverse effects on uses is done 3. Significantly affected uses based on the levels of significance are identified

Table 14.22. *Continued*

Country	Name of case study	Restoration measures to achieve GES	Positive effects of restoration measures	Costs of (setting up) restoration measures (not economic impact of measures)	Significant adverse effects on specified use? *Y(es)/N(o), Methods used*	Significant adverse effects on wider environment? *Y(es)/N(o), Methods used*	Remarks
		2. *Scenario B* – WB 1: provide weirs with fish ladders or bypass ways – WB 2: reduce volume of impounded lakes to avoid algal blooms			2. *Scenario B* – Reduction of hydropower generation in WB 2 – Limitation of passenger navigation in WB 2 – Limitation of maintenance navigation – Loss of area for agricultural land use	2. Construction of a bypass affords the digging of a new water course. Removed soil must be deposed filling the natural landscape	
	Mulde R.	1. Re-establish river continuum (remove/modify weirs) 2. Local redeployment or removal of dikes 3. Re-establish flow dynamics	1. Restore fish fauna 2. Former retention areas re-accessed		1. Problem to hydrpower plants due to long term permissions		
	Dhünn R.	Has not proceeded to designation					
E	Lozoya R.	1. Elimination of the dams (reservoirs) 2. Ecological water regime for the regulated rivers	1. Restoration of lotic conditions 2. Sufficient water regime for ichtyofauna		1. Y, loss of water supply for Madrid 2. – *qualitative assessment*		

Table 14.22. *Continued*

Country	Name of case study	Restoration measures to achieve GES	Positive effects of restoration measures	Costs of (setting up) restoration measures (not economic impact of measures)	Significant adverse effects on specified use? *Y(es)/N(o), Methods used*	Significant adverse effects on wider environment? *Y(es)/N(o), Methods used*	Remarks
SF	Kemijärvi L.	1. Lower minimum water level and decreased draw-down in winter	– Decrease in erosion of sandy shorelines – Recovering of frost sensitive species improvement of fish stock – Recreational use and fishing increased		1. Y, loss in hydropower production (30%)		
		2. Reproduction areas should be restored			2. Y, flood risk will significantly increase, damage for buildings and infrastructure		
F	Authie R.	1. Remove dams			1. It would lead to headward erosion (sinking of the riverbed)		
		2. Open sluice gates	2. Larger production area for migrating fish	2. Dam restoration costs should be borne by the dam owners/private individuals (French law) who all claim that they are unable to pay expenses	2. Y, economic loss due to the end of electricity production and of fish-breeding farms		
		3. Stop poplar tree plantation	3. Restore wetlands		3. Y, effect on the income of the owners of the plots near the rivers		
		4. Remove dikes downstream			4. Y, flood risk will rise		

Table 14.22. *Continued*

Country	Name of case study	Restoration measures to achieve GES	Positive effects of restoration measures	Costs of (setting up) restoration measures (not economic impact of measures)	Significant adverse effects on specified use? *Y(es)/N(o), Methods used*	Significant adverse effects on wider environment? *Y(es)/N(o), Methods used*	Remarks
		5. Stop extraction of granular particles			5. Y, loss of jobs (but permit of companies to extract particles is time limited)		
	Rhone R.	Improve migration continuity					
	Sarre R.	1. Removal of water intake (radical solution) or reserving a 2-year min. flow equivalent to 10-year natural flow (alternative solution)	1. Negligible impact on morphodynamics but of beneficial effect for connectivity and habitat	Costs calculated for each single measure proposed Total budget to reach GES at 4 625 000 Euros			Restoration measures defined by linking biological deterioration to physical modifications
		2. Removal of transverse constructions (radical solution) or replacement of dam with spillways or a series of submerged aprons or lowering of weirs (alternative solutions)			2. Erosive power of the flows may cut-off meanders and result in accelerated spread of flood waves (radical solution) Adverse effect on canalised and urbanised sections	2. Sudden lowering of water level due to incision of the bed may disturb adjoining wetland areas	

Table 14.22. *Continued*

Country	Name of case study	Restoration measures to achieve GES	Positive effects of restoration measures	Costs of (setting up) restoration measures (not economic impact of measures)	Significant adverse effects on specified use? *Y(es)/N(o), Methods used*	Significant adverse effects on wider environment? *Y(es)/N(o), Methods used*	Remarks
		3. Measures selected for the operation plan: Reserving min. flow equivalent to 10-year natural flow; removal/reduction of eddy zones; improving bank protection material; improving connectivity for fish migration					
GR	Nestos R.	*Scenario 1*: 'Maintaining current works': restoration of regulated flow, maintenance of ecological discharge, control of illegal irrigation canals (WB 3) improving irrigation network, maintenance of ecological discharge, restoring river meanders (WB 4) *Scenario 2*: 'Abandoning current works' removal of dams	Naturalised reservoir operation (WB 3) Maintenance of ecological discharge hinders groundwater salinisation Return to natural state		Halting reservoir operation is contrary to the beneficial objectives of the dams (WB 3) Energy and irrigation deficit Flooding of cultivations with a big impact on the rural population and the economy		

Table 14.22. *Continued*

Country	Name of case study	Restoration measures to achieve GES	Positive effects of restoration measures	Costs of (setting up) restoration measures (not economic impact of measures)	Significant adverse effects on specified use? *Y(es)/N(o), Methods used*	Significant adverse effects on wider environment? *Y(es)/N(o), Methods used*	Remarks
NL	Haringvliet Est.	*Scenario 1*: Removal of the dam, dikes remain, removal of bank reinforcement, dredging of contaminated sediments	– Restoration of estuarine salinity gradient – Restoration of tidal fluctuation		1. Y, reduced flood protection/safety		
		Scenario 2: Adjustment of the dam (lowering the bottom barrier), removal of bank reinforcement, dredging of contaminated sediments	– Restoration of estuarine morphological processes – Remediation of contaminated sediments		1 + 2.: Y: – Loss of agricultural water supply due to increased salinity ⇒ losses in yield of crops (6 billion Euros) – Loss of drinking water supply for 3 million inhabitants ⇒ *significant adverse effects* on specific uses (both scenarios) *semi-quantitative assessment*	1 + 2. Intertidal areas will partially disappear due to increased hydromorphological dynamics	
	Hagmolenbeek-Hegebeek R.	1. *Restoration of morphology* (meandering pattern, side-channels, removal of weirs)	1. Increased diversity of morphology and stream velocities, species diversity	1. Costs for buying land and remeandering the stream (1.5 million Euros) ⇒ high	1. N, loss of agricultural land bordering the stream, approx. 5% ⇒ not significant		Designation process assumes exact cause-effect-relationship between measure taken and resulting ecological condition, but it is often difficult to predict which measure would lead to GES

Table 14.22. *Continued*

Country	Name of case study	Restoration measures to achieve GES	Positive effects of restoration measures	Costs of (setting up) restoration measures (not economic impact of measures)	Significant adverse effects on specified use? *Y(es)/N(o), Methods used*	Significant adverse effects on wider environment? *Y(es)/N(o), Methods used*	Remarks
		2. *Restoration of the catchment,* (reconnection of former sub-catchments	2. Stimulation of rheophilic species	2. Construction costs for connection under the Twente canal (1.9 million Euros) ⇒ high	2. Water level increase of 1–20 cm, increase of flood risk ⇒ significance unclear	2. Streams in adjacent watersheds may run dry in dry periods or may be affected by ex-treme high discharges in wet periods	Prediction of effects of measure was difficult, lack of calibrated modelling approaches
		3. *Restoration of hydrology,* decreasing drainage, increase of groundwater level	3. Improvement of water quality due to longer retention of runoff		3. Y, increasing groundwater level affects agriculture: yield reduction, costs of yield reduction high (1.8 million Euros yr^{-1}) ⇒ significant *quantitative assessment*		
	Loosdrecht L.	1. Inundation of the southern part of the neighboured polders			1. Y, flooding of densely populated agricultural land, loss of drinking water supply for Amsterdam (30%)	1 + 2. Change of ground and surface water hydrology in the whole region	
		2. Rise of groundwater level			2. Y, effects neighbouring municipalities		
		3. Reduced water extraction in the infiltration area			3. Y, main drinking water source for the region Utrecht		

Table 14.22. *Continued*

Country	Name of case study	Restoration measures to achieve GES	Positive effects of restoration measures	Costs of (setting up) restoration measures (not economic impact of measures)	Significant adverse effects on specified use? *Y(es)/N(o), Methods used*	Significant adverse effects on wider environment? *Y(es)/N(o), Methods used*	Remarks
		4. Removal of unnatural embankments and yacht basins			4. Y, removal of 29 yacht basins, recreation function of the lake *qualitative assessment*		
	Veluwerandmeren	1. Natural water level management			1. Y, increased flood risk for towns and recreational facilities located at the border of the lakes	1. Y, water level management of Lake Ijssel has to change 30% of the Netherlands would have to adapt their water management	
		2. Removal of fortified banks, dikes, recreation beaches		2. Expensive for the Flevopolder side (3 m deep)	2. Y, removal of recreational infrastructure (yachting, camping) which is an important economic factor in the region		
		3. Restoration of the lake bottom, stopping mineral extraction			3. Y, shipping of goods will be impossible, mineral extraction will stop		
		4. Replacing seepage flushing by nutrient poor stream water, reduction of nutrient loads			4. Y, impact on agricultural yields *qualitative assessment*		

Table 14.22. *Continued*

Country	Name of case study	Restoration measures to achieve GES	Positive effects of restoration measures	Costs of (setting up) restoration measures (not economic impact of measures)	Significant adverse effects on specified use? *Y(es)/N(o), Methods used*	Significant adverse effects on wider environment? *Y(es)/N(o), Methods used*	Remarks
NO	Suldalslagen R.	1. Naturalised flow (release of 1350 million $m^3 a^{-1}$)	1.–3. Recovering of salmon population[a]		1. Y, 50% loss of HP production ⇒ significant	1. Increased flood damages, changed use of floodplains	[a] The assumption was made that the flow modification is responsible for salmon decline and not other reasons, such as acidification, parasites, and ocean pressures. Expert opinions differ on this
		2. *Restored flood regime* (occasionally large scouring floods of 500 $m^3 s^{-1}$ for 5 days)	1./2. Restoration of substrate, riverbed vegetation		2. N, loss of ca. 1% of HP production which can be absorbed in the energy production system ⇒ not significant	2. Flooding of floodplains; they would have to be adapted to the regulated flood regime ⇒ significant	
		3. Construction of fish passages	3. Migration continuity		3. – *quantitative assessment*	3. – *semi-quantitative assessment*	
	Beiarn R.	1. Reverse the water diversion			1. Y, reduced HP production (15%) *qualitative*		
S	Emån R.	1. Fish ladders	1. Migration continuity[b]	1. Building costs high			[b] Effect low since there are other obstacles downstream

Table 14.22. *Continued*

Country	Name of case study	Restoration measures to achieve GES	Positive effects of restoration measures	Costs of (setting up) restoration measures (not economic impact of measures)	Significant adverse effects on specified use? *Y(es)/N(o), Methods used*	Significant adverse effects on wider environment? *Y(es)/N(o), Methods used*	Remarks
		2. Minimum release of 30%	2./3. Improvement of riparian vegetation, improvement of fish habitats		2. Y, significant economic loss due to HP reduction (unprofitable)		
		3. Recreation of peak flows			3. N, no significant loss of HP production		
		4. Removal of HP stations			4. Y, loss of HP production *qualitative assessment*		
	Dalälven R.	1. Restoration of river continuum					
		2. Restoration of habitat diversity					
		3. Restoration of natural water flow			3. HP production would decrease		
		4. Restrict release of cultivated fish					
	Umealven R.	1. Restoration of natural flow patterns	Improved fish population, migration continuity		1.–3. Y, incompatible with large scale hydropower production (12% of annual hydropower production in Sweden) *qualitative assessment*		
		2. Restoration of transport of sediment, litter and organisms (removal of dam)					

Table 14.22. *Continued*

Country	Name of case study	Restoration measures to achieve GES	Positive effects of restoration measures	Costs of (setting up) restoration measures (not economic impact of measures)	Significant adverse effects on specified use? *Y(es)/N(o), Methods used*	Significant adverse effects on wider environment? *Y(es)/N(o), Methods used*	Remarks
		3. Restoration of rapids (removal of dam)					
	Baltic coast	1. Remove harbour facilities 2. Limitations on recreational shipping, infrastructure			1./2. Y, adverse effect on navigation, oil supply *qualitative assessment*		Reduction of local physical alteration is not enough to solve the env. Problems. Large scale effects on the whole Baltic Sea drainage must also be mitigated
UK (EW)	Kennet R. (Thames)	1. Removal of weirs with habitat restoration 2. Raise bed levels					The Kennet is un-natural and considerable management is required to maintain its status (also a site of special scientific interest)
	Tame R.	1. Take out flood banks 2. Remove bridges 3. Remove weirs	1. Improve spawning success 3. Favour upstream migration				

Table 14.22. *Continued*

Country	Name of case study	Restoration measures to achieve GES	Positive effects of restoration measures	Costs of (setting up) restoration measures (not economic impact of measures)	Significant adverse effects on specified use? *Y(es)/N(o), Methods used*	Significant adverse effects on wider environment? *Y(es)/N(o), Methods used*	Remarks
		4. Return to more natural channel	4. Favour colonisation of vegetation and improve fish larvae habitat				
	Sankey Brook (only Rainford Brook evaluated)	1. Raising riverbed level			1.–4. Unclear, loss of agricultural land, increase in flood damage		
		2. Channel narrowing					
		3. Removal of embanked sections					
		4. Re-engineering of river channel					
		5. Removal of culverts			5./6. Y, increase in mileage across all road users *qualitative assessment*		
		6. Removal of bridges				6. Increased air emission	
	Great Ouse R. (A) Main Ouse	1. Removal of all weirs, locks, reduction of channel width	Natural river channel, migration of salmonids		Y, loss of navigation link *qualitative assessment*	Potential impact to fens by saltwater intrusion	
	(B) Wissey	1. Removal of flood embankments	1.–4. Natural mosaic habitat, increase in limnophil fish, spawning areas		Y, loss of flood protection resulting in flooding of hundreds of km^2 of land situated lower than the river, comprising highly fertile agricultural land and housing *qualitative assessment*	Y, large scale flooding	

Table 14.22. *Continued*

Country	Name of case study	Restoration measures to achieve GES	Positive effects of restoration measures	Costs of (setting up) restoration measures (not economic impact of measures)	Significant adverse effects on specified use? *Y(es)/N(o), Methods used*	Significant adverse effects on wider environment? *Y(es)/N(o), Methods used*	Remarks
		2. Restoration of channel form					
		3. Connection to the lower part of Ouse					
		4. Removal of impoundment					
UK (S)	Forth Est.	*Water body B:*					
		1. Decommission of the harbour			1. Y, loss of main Scottish port for container ships, loss of industrial production (oil, chemical), loss of employment		
		2. Reverse land take, restore intertidal zone/ increase up to 70% of its original extent	2. Restoration of intertidal zone to 70% of its original extent		2. Y, loss of industrial production sites and agricultural lands, loss of employment		
		3. Restoration of intertidal areas taken for agriculture			3. Y, loss of highly valuable agricultural land (34%) *'Common sense', qualitative assessment*		
		Water body C:					
		1. Decommission of the Kincardine station and coal handling plant			1. Y, implications for the coal supply to Longanet Power station		

Table 14.22. *Continued*

Country	Name of case study	Restoration measures to achieve GES	Positive effects of restoration measures	Costs of (setting up) restoration measures (not economic impact of measures)	Significant adverse effects on specified use? *Y(es)/N(o), Methods used*	Significant adverse effects on wider environment? *Y(es)/N(o), Methods used*	Remarks
		2. Restoration of natural shoreline			2. Y, increased risk of flooding for industrial, commercial and residential property		
		3. Decommission of Longanet power station			3. Y, loss of power production associated with local coal mining, loss of employment		
	Tummel R.	1. Removal of all hydropower related structures and extensive restoration of streams, rivers and lochs	High ecological status Improvement of salmon population	1. Large cost (decommissioning of the dams, restoration works)	1. Y, loss of hydropower production, economic effects *qualitative assessment*	1. Loss of features valued for nature conservation and aesthetic reasons (reservoirs)	
	Dee R.	1. Appropriate compensation flows	1./3. Improvement of hydrological regime downstream	1./2. Capital cost of physical modifications to structures	1.–3. Y, economic effect: – loss of HP production (compensation water) – potential loss of revenue by imposing limitations on the power generation schedule that may prevent the operators from taking full advantage of the price incentives of peak-time provision *qualitative assessment*		Simply removing HP modifications/improving migration may not be sufficient to allow salmon recovery since acidity levels may still prevent recovery

Table 14.22. *Continued*

Country	Name of case study	Restoration measures to achieve GES	Positive effects of restoration measures	Costs of (setting up) restoration measures (not economic impact of measures)	Significant adverse effects on specified use? *Y(es)/N(o), Methods used*	Significant adverse effects on wider environment? *Y(es)/N(o), Methods used*	Remarks
		2. Fish passage and catchwater structures (arranging more suitable flows in the fish ladder to improve their effectiveness) 3. Complementary scheduling of electricity generation in the cascade to avoid simultaneous discharge into the WB 4. Removal of HP structures	2. Migration, fish habitat (though effect might be limited by acidification) 3. Improvement of hydrological regime downstream			4. Loss of wetland nature reserve downstream	
UK (NI)	Lagan R.	Many measures proposed associated with: – Weirs and blockages to continuity/fish passage – Changes to natural hydrological regime – Physical alteration to habitat – Changes to land use/ planning			Detailed assessment of each measure in proformas (see case study for detailed analysis). The methodology devel-oped by DEFRA (used in the UK-EW case studies) was used → restoration measures for WBs 4 and 5 have significant adverse effects. For WBs 3a and 3b, it is 'unsure' whether the restoration measures would have significant adverse effect and more detailed assessment is required		

Table 14.23. Case study results on Article 4(3)(b) – Assessment of other means (I). *Source:* Extracted from the case studies by the authors

Country	Name of case study	Benefits of existing use (modification)	"Other means"	Technical feasibility *Y(es)/N(o)*	Significantly better environmental option *Y(es)/N(o), Methods*	Remarks
A	Wienfluss R.	No information				
	Bregenzerach R.	Considered natural as of step 4(3)(a)				
	Danube R.		No other means (better env. options) were considered, since HP was considered as a comparatively environmentally friendly, renewable energy source			
B	Dender R.	1. Transport (navigation)	1. Road and train		1. N, external costs and envir. effects of different means of transport were compared	
		2. Income of the farmer (agriculture)	2. Other crops/forestry (causing less erosion)	2. Y, but social analysis may be needed (farmers not happy, insecure to change to forestry)		
		3. Flood protection	3. Construction of controlled floodplains	3. N, no suitable space	3. Y	
		4. Recreation	4. Soft recreation (angling, hiking)	4. N, in the strategic objectives for the Dender it is mentioned to promote recreational navigation	4. N, pressure on the environment will rise	
D	Elbe R.	Transport	1. Modification of navigation (local view)			

Table 14.23. *Continued*

Country	Name of case study	Benefits of existing use (modification)	"Other means"	Technical feasibility *Y(es)/N(o)*	Significantly better environmental option *Y(es)/N(o), Methods*	Remarks
			2. Replace with road or rail transport (regional view)		2. Y, comparison of ecological effects of the means of transport with regard to to the emission of pollutants, noise pollution demand for surface area → Railway transport has less environmental effects compared with inland navigation (related to energy consumption per transport unit)	
	Seefelder Aach R.	Agriculture: Income and employment for rural areas (within the EU)	Modification or abandonment of current land use structure; e.g. to diversify farm business in order to use other income possibilities	Y	Y, changes in land use patterns and abandonment of agriculture in floodplains will result in positive environmental effects in general. Restoration of river banks and floodplains will have effects on species and habitats, reduction in nutrient pollution in buffer strips etc. → Better environmental option	Assumption was made that land owners are willing to sell their properties
	Lahn R.	Power generation	Modification of HP generation		Estimation of avoidance costs: In relation to the total emissions in Germany 0.1% CO_2 emissions can be avoided due to hydro-power in general. For the small hydropower plants only, the percentage is accordingly lower	

Table 14.23. *Continued*

Country	Name of case study	Benefits of existing use (modification)	"Other means"	Technical feasibility *Y(es)/N(o)*	Significantly better environmental option *Y(es)/N(o), Methods*	Remarks
	Ruhr R.	1. Power generation (hydropower)	1. Wind power		1. Y, better or at least comparable environmental option	
		2. Recreation (passenger navigation)	2. Better fishing facilities and more attractive natural experiences		2. Y, better or at least comparable environmental option	
		3. Maintenance navigation (maintenance of the streambed for passenger navigation, protection of the waterworks and agricultural land)	3. (Scenario A): protection of waterworks without using ships		3. Cannot be decided	
		4. Agriculture (preservation of agricultural properties, conservation of landscape character)	4. Reactivation of abandoned agricultural land not in the floodplain Development of new agricultural land using the former lake ground		4. –	
	Mulde R.	Residential areas, energy production	Not possible to identify altermaives for the urban areas			
	Dhünn R.	Has not proceeded to designation				
E	Lozoya R.	1. Water supply	1. Water supply from other catchments, destruction of the dams	2. –	1. N, displacement of the impacts to other catchments, negative impacts from dam demolition work	

Table 14.23. *Continued*

Country	Name of case study	Benefits of existing use (modification)	"Other means"	Technical feasibility *Y(es)/N(o)*	Significantly better environmental option *Y(es)/N(o), Methods*	Remarks
		2. Hydropower	2. Ecological management of water flow from the reservoirs	2. Y	2. –	
SF	Kemijärvi L.	1. Flood protection	1. Increased storage capacity by constructing new reservoirs	1. Acceptance questionable	1. N, not a better env. option	
		2. Hydropower generation (30% of the Finnish short term capacity)	2. Dams and embankments in flood sensitive areas	2. Y	2. Y	
			3. Power production by other means (nuclear, wind, gas)	3. N, suitability for short term regulation (nuclear, wind)	3. N, fossil fuels and nuclear plants were not considered as better env. option	
F	Authie R.	Residential areas	Stop urbanisation	Y		
	Rhone R.	1. Transport	1. Sandbank restoration	Y	Y	
		2. Flood protection	2. Increase season-modulated water flow			
		3. Energy production	3. Minimum flow increase			
	Sarre R.	1. Flood protection 2. Residentlai areas 3. Transport	1. – 2. – 3. Modification of local infrastructure (roads and railways) + restoration measures proposed under test 4 (3)(a)	Y, only measures technically feasible were considered		

Table 14.23. *Continued*

Country	Name of case study	Benefits of existing use (modification)	"Other means"	Technical feasibility *Y(es)/N(o)*	Significantly better environmental option *Y(es)/N(o), Methods*	Remarks
GR	Nestos R.	Hydropower	1. Improvement of the irrigation network, water selection with smaller dams in tributaries			1. Cannot cover water demand
			2. Alternative energy production (gas)	2. N	2. Y	
NL	Haringvliet Est.	Flood protection Drinking water supply Agricultural water supply	1. Higher dike (maintains flood protection)+supplementary measures to maintain fresh water supply (other means not further assessed)			15 years were assumed as the time period for technical feasibility considerations. Further inclusion of social feasibility/acceptance was suggested
			2. Construction of a different dam + supplementary measures to maintain freshwater supply	2. Questionable (within 15 years)	2. Y, assumed that GES will be achieved	
	Hagmolenbeek-Hegebeek R.	Optimisation of hydrology for agricultural uses, increase of land area suitable for agriculture Reduced flood risk Navigation (Twentecanal)	1. Decreasing agricultural activity directly bordering the stream (±15 m)	1. – Technically feasible – Social feasibility is questionable. The area has to be bought on a voluntary basis, therefore the acceptance of measures by the local public and the farmers was questionable		It is difficult to determine the (technical) feasibility of certain measures and alternatives. Catchments and streams, which have been adapted to agricultural needs during centuries, cannot be restored to their natural condition within 15 years (timetable of the WFD).

Table 14.23. *Continued*

Country	Name of case study	Benefits of existing use (modification)	"Other means"	Technical feasibility *Y(es)/N(o)*	Significantly better environmental option *Y(es)/N(o), Methods*	Remarks
			2. Reduction of flood risk by creation of retention areas (about 3% of the catchment area)	2. Y		
			3. Increase of groundwater level	3. Technically feasible but it conflicts with Dutch policy of spatial planning; the catchment is assigned an agricultural function. Additionally, there are no alternative crops suitable for higher groundwater level		
	Loosdrecht L.	1. Flood protection	1. No alternative			
		2. Recreation	2. Displacement of the uses to another water body		2. N, displacement of the impacts	
		3. Drinking water extraction	3. Change drinking water supply from groundwater to surface waters	3. N, due to surface water pollution/salinity hyperfiltration is very expensive		
		4. Land for habitation and agriculture	4. Displacement of habitation and agriculture	4. N, high population pressure, socially not feasible	4. N, no env. better option	
	Veluwerandmeren	1. Flood protection	1. No alternative			
		2. Recreation	2. Displacement of the uses to another catchment		2. N, displacement of the problems	
		3. Shipping	3. Transport of goods by other means/other routes		3. Transport on roads not considered as a better env. option	

Table 14.23. *Continued*

Country	Name of case study	Benefits of existing use (modification)	"Other means"	Technical feasibility *Y(es)/N(o)*	Significantly better environmental option *Y(es)/N(o), Methods*	Remarks
		4. Mineral extraction	4. Stop of mineral extraction, mineral extraction on land	4. N, increasing need for mineral extraction; long term concessions have been made		
		5. Agriculture	5. Reduction/adaptation of agriculture			
NO	Suldalslagen R.	Hydropower production Recreation (salmon fishery) Flood protection	1. Naturalised flow and energy production by other means	1. Y (new production facilities might be necessary[a])	1. N, thermal power and nuclear power not considered as a better env. option	[a] Aggregation of many such restoration actions in the system would lead to the need for new energy production facilities, if it is not compensated with reduced releases in other HP stations where environmental consequences are less
			2. Restored flood regime	1./2. N, social feasibility due to flooding of floodplains is low Other means incompatible with present licensing operation	1./2. N, flood levee construction not considered as environmentally better option *qualitative assessment*	
	Beiarn R.	Hydropower production	1. Power production by other means/energy savings		1. Better env. option questionable (except for energy saving, reduction in heating by means in electricity)	
			2. Reverse diversion of water			
S	Emån R.	– Hydropower generation	1. Gradual increase/decrease when the release from HP plants is changed			

Table 14.23. *Continued*

Country	Name of case study	Benefits of existing use (modification)	"Other means"	Technical feasibility *Y(es)/N(o)*	Significantly better environmental option *Y(es)/N(o), Methods*	Remarks
		– Slow-flowing habitats/species	2. Minimum tapping of 5%	2. Y (no compensation necessary)		
		– High natural value of the reservoirs (diversity)	3. Energy produced by nuclear or fossil fuels		3. N, power production by fossil or nuclear no better option *qualitative assessment*	
	Dalälven R.	Energy production	1. Alternatives to hydropower		1. N, they cause other kinds of environmental damage	
			2. Better community house-keeping with electricity and heat			
	Umealven R.	Hydropower production (~2% of Swedish annual HP production)	1. Other power sources (nuclear and fossil fuels)	1. N, phasing out of nuclear power narrows the margin of reducing hydropower	1. No environmental better options without increasing costs of energy	
			2. Energy efficiency/saving and replacement of electricity for heating with other energy forms	2. It would require district heating	2. Y	
			3. Remedial measures (fish passages, compensation flow, flow management, naturalising of channels)		3. – *qualitative assessment*	
	Baltic coast	1. Harbour/oil supply	1. Reduced use of petroleum products			

Table 14.23. *Continued*

Country	Name of case study	Benefits of existing use (modification)	"Other means"	Technical feasibility *Y(es)/N(o)*	Significantly better environmental option *Y(es)/N(o), Methods*	Remarks
		2. Recreation	2. – Reduced speed of boats – Concentrate private small boats on marinas – Forbid boat engines in inlets			
UK (EW)	Kennet R. (Thames)	1. Flood protection	1. Enhancement of parallel channels and sluice systems Remove embankments/ more floodplain storage		1. Significant benefits to floodplain habitats	Detailed information in case study proforma tables
		2. Land drainage	2. No options			
	Tame R.		– Minor re-routing of the river (Stretch 1); – Removal of flood banks and provide a washland as protection to individual properties or provide a balancing lake (Stretch 2); – Remove flood banks (with or without protection works for individual properties) and/or railway bridges (Stretch 3); – Improve fish habitats by planting marginal vegetation (Stretch 5); – Modification of the river channel (Stretch 6); or	Detailed assessments of each measure in proformas (see case study)	Detailed assessments of each measure in proformas (see case study)	

Table 14.23. *Continued*

Country	Name of case study	Benefits of existing use (modification)	"Other means"	Technical feasibility *Y(es)/N(o)*	Significantly better environmental option *Y(es)/N(o), Methods*	Remarks
			– Introduction of wing dikes, backwaters or bays, or re-engineering of banks to increase diversity in and around the channel (Stretch 6) – Replacement of existing bank protection with softer (wooden) engineering solutions			
	Sankey Brook	Flood protection, agriculture	1. Parallel drainage channel	1. Y		
			2. Re-engineering of river channel, meanders	2. Y		
			3. Construction of new bridges	3. N, not feasible with constraints of road system		
	Great Ouse R. (A) Main Ouse	Navigation	1. Build a canal	1. Y (impractical/inconvenient)		
			2. Partly transport by road	2. Y	2. Questionable	
			3. Reduce width of channel	3. Y		
			4. Reprofiling of banks	4. Y		
			5. Adding non-navigable channels as 'natural habitats'	5. Y		
	(B) Wissey	Flood defence	No 'other means' to provide the flood defence function	N, no technically feasible alternatives to current modification		–

Table 14.23. *Continued*

Country	Name of case study	Benefits of existing use (modification)	"Other means"	Technical feasibility *Y(es)/N(o)*	Significantly better environmental option *Y(es)/N(o), Methods*	Remarks
UK (S)	Forth Est.	1. Navigation/harbour	WB B: 1. Remove harbour, transport by road/train	WB B: 1. Questionable	WB B: 1. N, increased road traffic, new road construction necessary	
		2. Land take for industrial purposes (power plant, chemical production)	WB B: 2. Remove power station, import of electricity from England WB C: 2. Remove docks		WB B: 2. N, higher SO_2 emissions, worse environmental option WB C: 2. – *qualitative assessment*	
		3. Land take for agriculture (livestock, crops, glass houses)	WB B: 3. Reverse agricultural reclamation WB C: 3. Reverse agricultural land take	WB B: 3. Y WB C: 1. Y	WB C: 1. Uncertain, depends on location of alternative production	
	Tummel R.	Hydropower production	1. Shut down HP operation and use other energy source	1. The alternative sources must be capable of meeting peak load demands at short notice/capacity in UK of fossil fuel power stations in long term run is limited	1. N, Fossil fuel power stations unlikely to be a better environmental option, due to increased CO_2 emissions and increased acid emissions	

Table 14.23. *Continued*

Country	Name of case study	Benefits of existing use (modification)	"Other means"	Technical feasibility *Y(es)/N(o)*	Significantly better environmental option *Y(es)/N(o), Methods*	Remarks
			2. Reduction of the environmental impacts whilst retaining some or all of the hydropower generation (increase compensation, fish passes, renaturalise flows)	2. Y		
	Dee R.	Hydropower production	1. Shut down HP operation and use other energy source	1. The alternative sources must be capable of meeting peak load demands at short notice/capacity in UK of fossil fuel power stations in long term run is limited	1. N, Fossil fuel power stations unlikely to be a better environmental option, due to increased CO_2 emissions and increased acid emissions (argument debatable/issue for national government)	
			2. Reduction of the environmental impacts whilst retaining some or all of the hydropower generation (increase compensation, fish passes, renaturalise flows)	2. Y	2. – *qualitative assessment*	
UK (NI)	Lagan R.		Many measures summarised in a table (see case study for details)	Detailed assessment of each measure in proformas (see case study for detailed analysis)	Detailed assessment of each measure in proformas (see case study for detailed analysis)	

Table 14.24. Case study results on Articl 4(3)(b) – Assessment of other means (II): disproportionate costs. *Source:* Extracted from the case studies by the authors

Country	Name of case study	Other means (alternatives) to the existing use	Existing use		Other means (alternatives)		Disproportionate costs? Results/Methods	Remarks
			Costs	Benefits	Costs (incl. fore-gone benefits)	Benefits (incl. benefits gained from higher ecological status)		
A	Wienfluss R. Bregenzerach R. Danube R.	No information Considered natural as of step 4(3)(a) No information						
B	Dender R.	1. Road and train transport (navigation) 2. Forestry/other crops (agriculture) 3. Controlled floodplains (flood protection) 4. Soft recreation (recreation)			2. High investment costs, profit on long term			
D	Elbe R.	1. Modification of navigation (local view)			No costs for 'navigation'. Foregone benefits cannot be assessed within this investigationt. It would require a detailed analysis of the operation costs, the freight charges and types of cargo ships		(use of absolute cost of alternatives) Costs are not disproportionate	When one considers the negative impacts of channelisation for navigation purposes for ecosystems and compares the low costs (income losses) and low profitability of inland navigation, the Elbe River should not be assessed as heavily modified

Table 14.24. *Continued*

Country	Name of case study	Other means (alternatives) to the existing use	Existing use		Other means (alternatives)		Disproportionate costs? Results/Methods	Remarks
			Costs	Benefits	Costs (incl. foregone benefits)	Benefits (incl. benefits gained from higher ecological status)		
		2. Replace with road or rail transport (regional view)						
	Seefelder Aach R.	Modification or abandonment of agriculture			Types of costs: – costs for land purchase and restoration measures: investment costs, implementation costs – operation and maintenance costs – opportunity costs (for possible forgone economic benefits) associated with financial losses for farmers due to lower yields on grassland or conversion of arable land into meadows		(use of absolute cost of alternatives) Costs are not disproportionate A cost-benefit analysis (CBA) would be appropriate but at the moment it is not yet possible to put a value on these benefits. Therefore, the decision was made only on cost-focused analysis comparing non-monetary benefits of restoration	The regional level for assessment is recommended in the case of agriculture A CBA is the best to meet the requirements during the designation process; it ensures that the alternative is chosen which provides a net benefit to society. A suitable and cost-effective approach might be the 'benefit transfer' as a method which uses the monetary values estimated at one study site and transposes them to another site. Costs of alternatives are better assessed with the ability to pay: total costs (for restoration as well as possible compensation payments) could be compared with the ability of local authorities to give support for the decision on the application of measures

Table 14.24. *Continued*

Country	Name of case study	Other means (alternatives) to the existing use	Existing use		Other means (alternatives)		Disproportionate costs? Results/Methods	Remarks
			Costs	Benefits	Costs (incl. foregone benefits)	Benefits (incl. benefits gained from higher ecological status)		
	Lahn R.	Modification of HP generation			Costs calculated; – costs for restoration of non-used weirs – investment costs – costs from foregone benefits		(use of absolute cost of alternatives) Costs are not disproportionate The production costs of plants which cannot reach break-even point indicate that power generation with small HP plants is a rather expensive way to avoid CO_2 emissions. Negative impacts (value of a restored water course) compensate the positive effects (avoidance costs)	In the case of small HP plants the primary question refers to the level considered, i.e. local or national alternatives. For the decision both views should be taken into account Local level: plants are not profitable in electricity generation National level: small HP plants are an expensive way to avoid CO_2 emissions

Table 14.24. *Continued*

Country	Name of case study	Other means (alternatives) to the existing use	Existing use		Other means (alternatives)		Disproportionate costs? Results/Methods	Remarks
			Costs	Benefits	Costs (incl. fore-gone benefits)	Benefits (incl. benefits gained from higher ecological status)		
	Ruhr R.	1. *Scenario A* (close to natural status) – abandon impounded lakes – replace weirs with artificial riffles – remove bank fixations locally – establish alluvial forests – reattach oxbow lakes			1. Calculation of replacement costs (lost profit): 1 610 175.91 Euros yr^{-1} Costs included total investment costs, total salary per year, running costs per year (operating costs, capital charges, maintenance costs)	1. Human benefit/positive effects (could not be calculated monetarily): – more attractive nature experiences – by abandoning the impounded lakes, former surface may be used for agriculture; eutrophication will be eliminated and costs of waste water treatment may be reduced; area of lakes may be used for spate retention; costs for sludge removal will be zero	1. (Comparison of costs with benefits of the alternative) *Scenario A*: Realisation is not a feasible aim (costs are higher than willingness to pay) *Method*: The costs to achieve GES are calculated. They include: – costs to replace existing uses (K_V) – costs realising alternatives to achieve the same benefit (K_A). – real costs to realise measures (K_P) Comparison of willingness to pay for nature protection (non use value) (Z_N) including expectable human benefit (use values) (W) with the estimated costs to achieve the GES	Three steps necessary: – identification of human benefit by achieving GES – calculation of costs to achieve GES and assessment whether they are proportionate – comparison with costs to achieve GEP Economic analysis was carried out for the restoration measures to achieve GES, not the 'other means'

Table 14.24. *Continued*

Country	Name of case study	Other means (alternatives) to the existing use	Existing use		Other means (alternatives)		Disproportionate costs? Results/Methods	Remarks
			Costs	Benefits	Costs (incl. foregone benefits)	Benefits (incl. benefits gained from higher ecological status)		
		2. *Scenario B* – WB 1: provide weirs with fish ladders or bypass ways – WB 2: reduce volume of impounded lakes to avoid algal blooms			2. For WB 1: willingness to pay, use values and costs (to replace existing uses, to realise alternatives and other investments) were calculated for 15 years. Benefit-cost-quotient was 1.24, therefore willingness to pay is higher than costs For WB 2, the following were calculated: actual profit per year, future profit after lowering the water level, costs for the production of energy by wind engines per year. The benefit-cost-quotient was 0.44, therefore willingness to pay is lower than costs	2. Human benefit/positive effects: – bypass streams will increase attractivity of landscape – area can be used for spate retention	The costs are considered proportionate if the benefit-cost-quotient is at least equal to 1: $Z_N + W / K_V + K_A + K_P > 1$ 2. (Comparison of costs with benefits of the alternative) *Scenario B*: willingness to pay is higher than costs in WB 1. Costs are higher than willingness to pay in WB 2	

Table 14.24. *Continued*

Country	Name of case study	Other means (alternatives) to the existing use	Existing use		Other means (alternatives)		Disproportionate costs? Results/Methods	Remarks
			Costs	Benefits	Costs (incl. fore-gone benefits)	Benefits (incl. benefits gained from higher ecological status)		
	Mulde R.	No information						
	Dhünn R.	WB has not proceeded to designation						
E	Lozoya R.	1. Water supply from other catchments, destruction of the dams 2. Ecological management of water flow from the reservoirs					No economic evaluation	
SF	Kemijärvi L.	1. Increase storage capacity by constructing new reservoirs 2. Dams and embankments in flood sensitive areas 3. Power production by other means (nuclear, wind, gas)				Benefits from recreation and fishing estimated (0.1–0.5 million Euros yr^{-1})	2. Disproportionate costly for this specific use (summer houses)	
F	Authie R.	Stop urbanisation					No information	

Table 14.24. *Continued*

Country	Name of case study	Other means (alternatives) to the existing use	Existing use		Other means (alternatives)		Disproportionate costs? Results/Methods	Remarks
			Costs	Benefits	Costs (incl. fore-gone benefits)	Benefits (incl. benefits gained from higher ecological status)		
	Rhone R.	1. Sandbank restoration					Measures are of reasonable cost	However, water bodies are designated as HMWB because proposed measures are not enough to restore fully active hydrological performance
		2. Increase season-modulated water flow						
		3. Minimum flow increase						
	Sarre R.	Modification of local infrastructure (roads and railways) + restoration measures proposed under test 4(3)(a)					Not possible given the socioeconomic and budgetary capacity of the Sarre basin	
GR	Nestos R.	1. Improvement of the irrigation network, water selection with smaller dams in tributaries 2. Alternative energy production (gas)					No economic evaluation (no technically feasible or economically acceptable alternatives, alternative energy production is too expensive)	

Table 14.24. *Continued*

Country	Name of case study	Other means (alternatives) to the existing use	Existing use		Other means (alternatives)		Disproportionate costs? Results/Methods	Remarks
			Costs	Benefits	Costs (incl. fore-gone benefits)	Benefits (incl. benefits gained from higher ecological status)		
NL	Haringvliet Est.	1. Higher dikes (maintains flood protection) + supplementary measures to maintain fresh water supply				Ecological benefits not quantified/monetised	(use of absolute costs of alternative) Absolute costs disprop. ⇒ 1 400 million Euros	Costs cannot be distin-uished between the 3 WBs [a] It remained uncertain where and when to consider costs of sediment remediation. Should they be considered as part of the RBMP or within the costs analysis of Art. 4 (3)(b) assuming it as new threat due to restoration measures? Economic assessment was carried out only for the most realistic option after qualitative consideration
		2. Construction of a different dam + supplementary measures to maintain freshwater supply			2. Destruction of the dam, capital costs for a new dam, freshwater supply Costs for sediment remediation[a]			

Table 14.24. *Continued*

Country	Name of case study	Other means (alternatives) to the existing use	Existing use		Other means (alternatives)		Disproportionate costs? Results/Methods	Remarks
			Costs	Benefits	Costs (incl. fore-gone benefits)	Benefits (incl. benefits gained from higher ecological status)		
	Hagmolenbeek-Hegebeek R.	1. Decreasing agri-cultural activity directly bordering the stream	1. Costs of present maintenance (mowing, clearance)	Not considered	1. Buying of land, construction costs	Not considered	(comparison of costs of alternatives) 1. Comparison of maintenance costs ⇒ alternative 65% lower	
		2. Reduction of flood risk by creation of retention areas (about 3% of the catchment area)			2. Costs for construction of connection, compensation costs for retention areas			
		3. Increase in groundwater level			3. Costs of yield reduction (1.8 million Euros yr^{-1}) Maintenance costs			
	Loosdrecht L.	2. Displacement of the uses to another WB 3. Change drinking water supply from groundwater to surface waters 4. Displacement of habitation and agriculture					No economic evaluation	

Table 14.24. *Continued*

Country	Name of case study	Other means (alternatives) to the existing use	Existing use		Other means (alternatives)		Disproportionate costs? Results/Methods	Remarks
			Costs	Benefits	Costs (incl. fore-gone benefits)	Benefits (incl. benefits gained from higher ecological status)		
	Veluwerandmeren	1. Displacement of the use to another catchment 2. Transport of good by other means/other routes 3. Stop of mineral extraction, mineral extraction on land 4. Reduction/adaption of agriculture			1. Loss of labour and income		1. ⇒ disproportionate *qualitative*	
NO	Suldalslagen R.	1. Naturalised flow and energy production by other plants 2. Restored flood regime	Not considered	1. Turnover from power production, salmon fisheries and recreation 2. Turnover from power production, salmon fisheries and recreation	1. Reduction in power production, costs from flood damages 2. Reduction in power production, costs from flood damages (non-verified estimates)	1. Increased turnover from recreation and salmon fisheries 2. 2 × higher salmon production, benefit for recreation (non-verified estimates)	1. Comparison of change in production and turnover (power, salmon fisheries, recreation and floodplains) ⇒ negative net-benefit 2. Comparison of change in production and turnover (power, salmon, floodplains) ⇒ positive net-benefit	

Table 14.24. *Continued*

Country	Name of case study	Other means (alternatives) to the existing use	Existing use		Other means (alternatives)		Disproportionate costs? Results/Methods	Remarks
			Costs	Benefits	Costs (incl. fore-gone benefits)	Benefits (incl. benefits gained from higher ecological status)		
	Beiarn R.	1. Power production by other means	Not considered	Turnover, Power production		No benefits	Comparison of turnover, income of existing use/alternative ⇒ disproportionate (negative net-benefit)	
		2. Reverse diversion of water (reduction of HP by 15%)			2. Reduction in power production			
S	Emån R.	1. Minimum tapping of 5%	Not considered	Annual power production	1. 5% reduction of HP production	Value of fish stock measured by 'willingness to pay for recreational fishing'	1. Gained ecol. benefit 5 times higher than loss → *net benefit*, no disprop. costs	Hard to estimate the economic value of species/nature
		2. Gradual increase/decrease when the release from HP plants is changed						
		3. Energy produced by nuclear of fossil fuels						
	Dalälven R.		No information					
	Umealven R.	1. Other power sources (nuclear and fossil fuels)	Not evaluated	Not evaluated	Not evaluated	Not evaluated	1. Disproportionate costs, increase in electricity prices	

Table 14.24. *Continued*

Country	Name of case study	Other means (alternatives) to the existing use	Existing use		Other means (alternatives)		Disproportionate costs? Results/Methods	Remarks
			Costs	Benefits	Costs (incl. fore-gone benefits)	Benefits (incl. benefits gained from higher ecological status)		
		2. Energy efficiency/saving Replace electricity for heating with other energy forms 3. Remedial measures (fish passages, compensation flow, flow management, naturalising of channels)					2. Not disprop. costly 3. Probably disproportionate costs *qualitative assessment*	
	Baltic coast	1. Reduced use of petroleum products 2. Reduced speed of boats concentrate private small boats on marinas, forbid boad engines in inlets					No economic evaluation	
UK (EW)	Kennet R. (Thames)	– Enhancement of parallel channels and sluice systems – Remove embankments/more floodplain storage					Calculation of operating and maintenance costs of existing modification; capital and maintenance/operating costs of alternatives;	Detailed information in case study proforma tables

Table 14.24. *Continued*

Country	Name of case study	Other means (alternatives) to the existing use	Existing use		Other means (alternatives)		Disproportionate costs? Results/Methods	Remarks
			Costs	Benefits	Costs (incl. fore-gone benefits)	Benefits (incl. benefits gained from higher ecological status)		
							potential environmental benefits – benefit/cost transfer assessment In overall, rehabilitation costs would be high. Because benefits would be extensive, further quantification of both is necessary	
	Tame R.	1. Remove floodbanks and provide a washland upstream in Stretch 2 2. Leave bridges, remove flood banks in Stretch 3 3. Planting marginal vegetation in Stretch 5 4. Raising of the riverbed, introduction of riffles or shoals, remove unnatural bank profiles in Stretch 6					Calculation of operating and maintenance costs of existing modification; capital and maintenance/operating costs of alternatives; potential environmental benefits – benefit/cost transfer assessment All costs are given in Present Value terms (i.e. discounted back to year 0 using a discount rate of 6%). This means they can then be compared with benefits accrued from the measures	Detailed assessment of each measure in proformas (see case study for detailed analysis)

Table 14.24. *Continued*

Country	Name of case study	Other means (alternatives) to the existing use	Existing use		Other means (alternatives)		Disproportionate costs? Results/Methods	Remarks
			Costs	Benefits	Costs (incl. fore-gone benefits)	Benefits (incl. benefits gained from higher ecological status)		
	Sarkey Brook						Calculation of operating and maintenance costs of existing modification; capital and maintenance/operating costs of alternatives; potential environmental benefits – benefit/cost transfer assessment	Detailed assessment of each measure in proformas (see case study for detailed analysis)
	Great Ouse R. (A) Main Ouse		Costs of operation and maintenance <1.6 million Euros	Not considered	Capital costs for construction, operation and maintenance		Calculation of operating and maintenance costs of existing modification; capital and maintenance/operating costs of alternatives; potential environmental benefits – benefit/cost transfer assessment ⇒ No disprop. costs	Detailed assessment of each measure in proformas (see case study for detailed analysis)
		1. Build a canal 2. Partly transport by road 3. Reduce width of channel			1.–4. >16 million Euros ⇒ disprop. cost			

Table 14.24. *Continued*

Country	Name of case study	Other means (alternatives) to the existing use	Existing use		Other means (alternatives)		Disproportionate costs? Results/Methods	Remarks
			Costs	Benefits	Costs (incl. fore-gone benefits)	Benefits (incl. benefits gained from higher ecological status)		
		4. Reprofiling of banks 5. Adding non-navigable channels as 'natural habitats'			5. <16 million Euros ⇒ no disprop. costs	5. Benefits of angling estimated Impossible to value water quality, physical habitat, conservation value, flora	5. Costs versus benefits from angling ⇒ net-benefit, *not disprop. costly*	
	(B) Wissey	No alternatives						
UK (S)	Forth Est.	*WB B:* 1. Remove harbour, transport by road/train 2. Remove power station, import of electricity from England	Not considered	1. Value of industrial output (500 million Euros yr^{-1}) Jobs considered for harbour, power station and coal mine	*WB B:* 1. Extra cost for road transportation (million Euros yr^{-1}) 2. Annual value of lost output (50 million Euros), effects on mining	Monetary evaluation of benefits from ecological restoration is not possible due to an absence of any suitable studies from which values can be transferred No quantification of benefit possible	*WB B:* 1./2. *Disprop. costs*	

Table 14.24. *Continued*

Country	Name of case study	Other means (alternatives) to the existing use	Existing use		Other means (alternatives)		Disproportionate costs? Results/Methods	Remarks
			Costs	Benefits	Costs (incl. fore-gone benefits)	Benefits (incl. benefits gained from higher ecological status)		
		3. Reversing agricultural reclamation			3. Net social value of annual lost farm output			
		WB C:			*WB C:*		*WB C:*	
		1. Reversing agricultural land take			1. Net social value of lost annual farm output		1. Absolute cost per household in the area not too high: ⇒ *not disprop. costly*	
		2. Removing of docks			2. Not able to quantify costs of rebuilding of properties liable to flooding			
	Tummel R.	1. Shut down HP operation and use other energy source	Not considered	Market value of electricity	1. Costs of decommissioning dams and structures, construction of fossil fuel power plants ⇒ *large costs*		1. Absolute costs high	No attempt has been made to quantify benefits to the ecology, which would require a system to express ecological benefits in monetary terms
		2. Reduction of the environmental impacts whilst retaining some or all of the HP generation (increase compensation, fish passes, renaturalise flows)			2. Annualised capital and running costs + loss of power generation revenue These are calculated for each restoration measure	2. Welfare measures of salmon fishing (see Dee River) ⇒ lower estimate, other ecological benefits not estimated	2. Cost alternative compared to gained ecological benefit for each measure ⇒ negative net-benefit, disprop. costly	

Table 14.24. *Continued*

Country	Name of case study	Other means (alternatives) to the existing use	Existing use		Other means (alternatives)		Disproportionate costs? Results/Methods	Remarks
			Costs	Benefits	Costs (incl. fore-gone benefits)	Benefits (incl. benefits gained from higher ecological status)		
	Dee R.	1. Shut down HP operation/other energy source 2. Reduction of the environmental impacts (fish passes, compensation flows)	Not considered	Market value of electricity	1. (a) Costs of decommissioning dams and structures (b) Construction of fossil fuel power plants 2. Annualised capital and running costs and loss of power generation revenue Calculated for each restoration measure in each WB	2. Ecological benefits Welfare measures of salmon fishing, (predicted by the change in salmon fishing in the affected areas, valuing these using current rents for privately-owned beats, based on an average value per km) ⇒ lower estimate, other ecological benefits not estimated	1. Absolute costs large 2. Cost alternative compared to gained ecological benefit for each measure and for each WB ⇒ for 4 WBs, the net-benefit was positive ⇒ for 5 WBs, the net-benefit was negative	Lack of studies to place adequate monetary values on ecological improvements resulting from restoration measures
UK (NI)	Lagan R.	Long list of measures summarised in a table (see case study for details)					Calculation of operating and maintenance costs of existing modification; capital and maintenance/operating costs of alternatives; potential environmental benefits	Detailed assessment of each measure in proformas (see case study for detailed analysis)

Table 14.25. Mitigation measures for GEP and MEP with respect to sustainability of fish populations in the Dalälven River (S). *Source:* Case study on the River Dalälven (S)

	Type of criteria	Target unit	Ecological potential	
			Good	Maximum
1	River continuum	*River catchment*	Well functioning passages to and from the sea	Highly functioning passages to and from the sea
2	Degree of fragmentation	*Main river or catchment >100 km²*	<25%	<10%
3	Degree of exploitation of hydropower potential (height or effect)	*Catchment*	<75%	<50%
4	Barrier effect			
	% of potential reach with reproduction areas	*Catchment*	>30%	>50%
5	Effects on flow regime			
	Natural annual variation (spring and autumn floods)	*River*	>75% level	>75% level and natural duration
	Water discharge (% of natural)	*River*	15%	30%
	Proportion of natural stream reaches (%)	*River*	>50%	>75%
6	Type of regulation			
	Dry reach, permanent	*Downstream dam*	Impossible	Impossible
	Dry reach, minimum discharge (% of natural)	*Downstream dam*	At least 15% of natural discharge	At least 30% of natural discharge
	Short-term fluctuations, weekly	*Run-off-river impoundment*	At least 30% of natural discharge	Impossible
	Short-term fluctuations, daily	*Run-off-river impoundment*	At least 30% of natural discharge	Impossible
	Short-term fluctuations, hourly	*Run-off-river impoundment*	Impossible	Impossible
7	Regulation amplitude	*Reservoirs*	4 m	2 m
		Regulated lakes	2 m	1 m
8	Possibilities for fish migration		Kaplan turbines with well functioning fish passages	Highly functioning fish passages governed after fish migrations
	Turbine type	*Hydropower dam*	Kaplan	Kaplan
	Mechanical leading of fish away from turbine intake	*Hydropower dam*	Well functioning	Highly functioning
	Bypass for fish	*Hydropower dam*	Well functioning	Highly functioning
	Fish ladders or bypass	*Dam*	Well functioning	Highly functioning

Table 14.25. *Continued*

Type of criteria		Target unit	Ecological potential	
			Good	Maximum
9	Canalisation			
	Stream bed cleared of stream thresholds, rocks, large stones etc. (former logging passages)	*Reach longer than 1 km*	Restoration of 75% of stream/riverbed structure to nature-like conditions for each km	Restoration of natural stream/riverbed structure to nature-like conditions
	Channelled with artificial beach/quay	*Reach longer than 1 km*	Suitable measure to increase habitat diversity for each km	Restoration of natural meandering

Table 14.26. Case study methods for the definition of MEP (I). *Source:* Extracted from the case studies by the authors

Country	Name of case study	Hydromorphological (A) and physicochemical (B) element	Modification necessary to support the use	Mitigation measures (including best environmental practice)	Ecological effects of mitigation measures (mainly on biota)	Comments – Remarks
A	Wienfluss R.	A: Constantly flowing main arm and several more or less well linked side arms Floods leading to substrate shifts, creating new structures and pioneer sites		– Restitution of far-reaching run-off dynamics, corresponding to the natural discharge regime – Improvement of the connectivity between Danube, Donaukanal and Wienfluss	Enable fish migration	
	Bregenzerach R.	Threshold values for the Bregenzerach would be 1:3 for the maximum daily flow amplitude and 20% maximum daily flow alteration within 15 min	Hydropower plant (hydropeaking effects)	Keep flow amplitudes and sudden flow increases within natural flow fluctuations		
	Danube R.		Impoundment	The definition of mitigation measures is based on state-of-the-art concepts and technologies keeping in mind that regular adaptations may be required depending on current developments Mitigation measures: – 1/3 lowering of water level in the impoundment to create a free flowing section – Creating dynamic floodplains along the re-established free flowing section – Reconstructing the mouth of the Gießgang – Re-establishment of longitudinal and lateral connectivity – Structuring the river bottom and banks along the impoundment compatible with navigation		The appropriate level of mitigation has been defined by expert judgement based on experiences of a variety of restoration activities in Austrian lowland rivers and at the Danube

Table 14.26. *Continued*

Country	Name of case study	Hydromorphological (A) and physicochemical (B) elements	Modification necessary to support the use	Mitigation measures (including best environmental practice)	Ecological effects of mitigation measures (mainly on biota)	Comments – Remarks
B	Dender R.		Locks	1. Reconnecting old meanders 2. Reconstruction of the embankments (fractured stones) 3. Placing of deflectors to increase structural diversity 4. Constructing fish passages or diversion channels 5. Introducing a controlled reduced tide 6. Erosion control measures	1. Develop macrophytes 2. Growth of aquatic plants 3. Increase fish population, raise macrophytes variation 4. Improve fish migration	Measures with negative effects on uses or technically infeasible measures were not considered
D	Elbe R.	No information				
	Seefelder Aach R.	No information				
	Lahn R.	No information				
	Ruhr R.	Not mentioned		Not mentioned		
	Mulde R.	Definition of MEP not feasible. The ecological status of the study area is dominated by its chemical status				
	Dhünn R.	WB not HMWB				
E	Lozoya R.	A: Residence time, depth, water level fluctuations Morphological: sediment structure, shape of shores B: Physicochemical quality elements, nutrients	Dam (reservoirs)	*Measures to improve hydromorphol. quality* 1. Minimum water level fluctuations compatible with the current water supply 2. Creation of wetlands in the riparian zone of reservoirs 3. Regeneration of the littoral zone 4. Fish passages *Measures to improve physicochemical quality* 5. Reduction of diffuse source and point discharges: creation of wetlands, waste water treatment facilities	1.–3. Continuity with terrestrial ecosystems, improving habitat structure 4. Improving fish community, migration continuity 5. Improvement of the trophic status	

Table 14.26. *Continued*

Country	Name of case study	Hydromorphological (A) and physicochemical (B) elements	Modification necessary to support the use	Mitigation measures (including best environmental practice)	Ecological effects of mitigation measures (mainly on biota)	Comments – Remarks
SF	Kemijärvi L.	A: Water level fluctuation	Water level regulations, sequence of hydropower stations downstream	– Sight or moderate changes in current regulation practice – Restoration of important bird areas – Restoration of tributaries – Bottom weirs in sheltered bays – Fish ladders – Fish stockings	Positive impacts on littoral ecosystem and fish stocks (predicted by modelling (REGCEL water level analysis tool) – Refuge for ice sensitive species – Improvement of spawning grounds	Effects of fish ladders questionable since there are several power stations in sequence
F	Authie R.	Not mentioned		– Reduction of impact of diffuse pollution – Co-ordination of opening of floodgates/active awareness of the public		
	Rhone R.		Canalisation, modifications for urbanisation			Mitigation measures for MEP should be tested for technical feasibility, economic environment, financial incentives and statutory regulations
	Sarre R.					
GR	Nestos R.	No information				
NL	Haringvliet Est.	A: Tidal amplitude, maximum current velocity Values for hydromorphological quality elements at MEP were based on model calculations B: No particular reference water body was selected to	Dam	1. Management of the sluices as storm surge barrier, relocation of drinking water inlet points 2. Remediation of contaminated sediments 3. Partial removal of shore protection works	– Restoration of the former estuarine characteristics, recovery of brackish nature values – Free migration of organisms – Reestablishment of marine organisms Effect of measures on biology was predicted by modelling and expert judgement	Only technically feasible measures were considered. Open question remained whether a cost analysis on the disproportionality of the measures for MEP should be carried out. Or should the question of costs be considered in the RBMP?

Table 14.26. *Continued*

Country	Name of case study	Hydromorphological (A) and physicochemical (B) elements	Modification necessary to support the use	Mitigation measures (including best environmental practice)	Ecological effects of mitigation measures (mainly on biota)	Comments – Remarks
		provide quantitative values for physicochemical elements. Instead, the generic target values used in the Netherlands were applied				There is little information about the interaction between water quality, habitat quality and the presence and functioning of species. It is therefore difficult to define the way biology will respond to the mitigation measures presented for MEP or GEP
	Hagmolenbeek-Hegebeek R.	A: Quantity of flow, flow dynamics, river depth width, substrate streambed B: MEP values for nutrients, oxygen based on regional standards of negligible risk	River straightening, channelisation	– Stream restoration (re-meandering) – Restoration of the catchment – All mitigation measures that have to be taken to ensure the best approximation to ecological continuum – Measure to achieve the water quality objectives (feasibility of measures evaluated in a table)		
	Loosdrecht L.	B: Physicochemical parameters (T, O_2, phosphate, transparency)	Controlled water level, bank fixation	1. Reduction of P load (dephosphorization of seepage; biomnanipulation) 2. Decreasing of turbidity (creation of silt catch) 3. Removing of bank fixations, creating natural banks and marshes	1. De-eutrophication 2. Ecological effect of the silt catch remains uncertain 3. Habitat improvement	

Table 14.26. *Continued*

Country	Name of case study	Hydromorphological (A) and physicochemical (B) element	Modification necessary to support the use	Mitigation measures (including best environmental practice)	Ecological effects of mitigation measures (mainly on biota)	Comments – Remarks
	Veluwerandmeren	B: Physicochemical parameters (T, O_2, phosphate, transparency	Controlled water level, bank fixation	1. Creation of natural banks and marshes 2. Reduction of nutrient load of incoming water (sewage works)	1. Increasing ecological quality of the banks 2. Stabilisation of the current status	The measures may conflict with the Birds Directive. The unnatural low water levels especially in winter as well as the high biomass of fish lead to a high density of birds. It cannot be excluded that the restoration measure may lead to a decrease of the diversity and abundance of birds which is in conflict with the Birds Directive
NO	Suldalslagen R.	A: Flow regime B: Physicochemical elements: (pH, T, nutrients), synthetic and non synthetic pollutants	Dams (hydropower)	– Restored flood regime – Reducing erosion in tributaries – Instream habitat improvement – Build fish passages	Improved ecological continuum	
	Beiarn R.	A: Flow	Water abstraction via tunnel intakes	Ensure stable river flow below the intakes, best operation practice	Ecological continuum will not be reached (interrupted by tunnel intakes)	

Table 14.26. *Continued*

Country	Name of case study	Hydromorphological (A) and physicochemical (B) elements	Modification necessary to support the use	Mitigation measures (including best environmental practice)	Ecological effects of mitigation measures (mainly on biota)	Comments – Remarks
	Emån R.		Dams (hydropower)	– Minimum release of 5% – Gradually increase/decrease in tapping – Restoration of cleared and straightened stretches – Fish ladders – Creation of peak flows	– Improved habitats in dry river channels at the hydropower plant (not enough to achieve viable fish stock) – Reduced flow stress on fauna especially macroinvertebrates – Creation of river spawning grounds – Ecological continuum – Create spatial and temporal fluctuation, increase in biodiversity	Research on the effects of the suggested mitigation measures is needed in order to make a good approximation of MEP
	Dalälven R.	A: River continuum, degree of fragmentation, flow regime	Dams (hydropower)	Measures defined from literature and expert judgement – Building well-functioning fish passages – Reduce the storage of water for hydropower to gain more natural flow regime		Measures which would not cause 'adverse effect upon uses' should be considered
	Umealven R.	A: Flow	Dams (hydropower)	– Ecological water flow and water level fluctuations (interferes with power production, production losses up to 10% assumed as acceptable) – Fish passages – Habitat improvements (boulders, erosion reduction) – Co-ordinated spillway releases – Seed and plant species – Removal of plant biomass	Increased biodiversity, habitat renewal, sedimentary processes, ecological continuum, habitat improvement, recovering of natural riparian and aquatic plant community	Knowledge of response to mitigation measures, difficult to define exact values for MEP

Table 14.26. *Continued*

Country	Name of case study	Hydromorphological (A) and physicochemical (B) elements	Modification necessary to support the use	Mitigation measures (including best environmental practice)	Ecological effects of mitigation measures (mainly on biota)	Comments – Remarks
	Baltic coast	A: Hydromorphological quality elements B: Physico-chemical quality elements	Navigation and recreation facilities	Nynäshamn: Cease the discharge of petrochemical compounds, restriction of harbour activities, stricter environmental standards for boats, reduction of nitrogen and phosphorus discharges Landfjärden: Regulations for recreational crafts (speed…), restrictions for the construction of new recreational cottages, control of private wastewater treatments, regulation of sport fishing in spawning periods		
UK (EW)	Kennet R. (Thames) Tame R. Sankey Brook Great Ouse R. (A) Main Ouse (B) Wissey			(B) 1. Restoration of wetlands 2. Restoration of marginal habitat 3. Return to 'natural' channel	1. Favour limnophil fish species 2. Colonisation by marginal vegetation useful as refuge habitat for fish larvae and spawning habitat 3. Improve colonisation or marginal vegetation and improve survival of coarse fish larvae and young	

Table 14.26. *Continued*

Country	Name of case study	Hydromorphological (A) and physicochemical (B) elements	Modification necessary to support the use	Mitigation measures (including best environmental practice)	Ecological effects of mitigation measures (mainly on biota)	Comments – Remarks
UK (S)	Forth Est.	A: Morphology of shores	Land claim Dredging Power station	Minimise dispersion of dredged spoil outwith the designated channel (best environmental practice) Reduce rate of water intake at power station to typical levels for 'dry' technology (best environmental practice) Reduce temperature of water discharged from station to ambient Re-establish estuarine habitats	Positive effects (increase in abundance and biodiversity) for angiosperms, benthic invertebrate fauna, fish	
	Tummel R.	A: Flow, sediment B: Temperature	Dams and weirs	– Fish passages – Reverse acidification (liming) – Import of appropriate sediment[a] – Compensation water at ambient temperature[a] In some WBs compensation flows and modified operation schedule	– Ecological continuum (migratory fish) – Increase in acid sensitive species – Increased survival of juvenile salmonids – Increased benthic invertebrate fauna – Increased spawning density – Reduction of algal abundance – Re-synchronising of spawning and juvenile growth	[a] Further assessment on impact, practicability and effects necessary

Table 14.26. *Continued*

Country	Name of case study	Hydromorphological (A) and physicochemical (B) elements	Modification necessary to support the use	Mitigation measures (including best environmental practice)	Ecological effects of mitigation measures (mainly on biota)	Comments – Remarks
	Dee R.	A: Hydrological regime, river continuity, morphological conditions B: General, synthetic and non-synthetic pollutants (Potential general conditions for MEP should incorporate theoretical resolution of acidification as well as other chemical problems. If GEP cannot be reached due to unavoidable impacts, less stringent environmental objectives may be set. These must represent the highest ecological and chemical status possible given the unavoidable impact.)	Dams, catch-water intakes for electricity generation	A code of best practice is needed on appropriate mitigation measures to counteract sediment starvation downstream of dams and catchwater intakes. Measures were proposed building on elements of good practice established by the HP operators Measures were proposed for the separate groups of water bodies. Some mentioned are: – facilities for fish passage; – channel renaturalisation; – reversal of acidification; – draw water from epilimnion; – import appropriate sediment; – reduce organic enrichment below fish farms; – provision of feasible compensation flow; and, – adjust timing of HP generation to achieve more equitable flow regime	Potential effects on biological quality elements in the individual water bodies were assessed based on expert judgement	Mitigation measures considered should not have a significant adverse effect on the use Preferred measures were those that simply involve changes in existing practice. Measures with capital cost implications and non-significant effects on the use were also considered Many measures remain subject to considerable uncertainty; further research is required to establish quantitative responses
UK (NI)	Lagan R.		Weirs, canalisation	Removal of weirs, changes in bridge construction, reduction of canalisation, reinstatement of riparian buffer zone (WB 3) Reduction of input in nutrients (WB 4 and 5) Direct aeration of the water column, flushing of the impoundment with sea water, periodic dredging over winter instead of large dredging operations (WB 4)		To define MEP and mitigation measures, advice is needed on the effects of the current modifications in order to be consistently estimated and interpreted. Pollution pressures confound the effects of modifications

Table 14.27. Case study method for the definition of MEP (II). *Source:* Extracted from the case studies by the authors

Country	Name of case study	Biological elements and methods used to derive their values	Results on the determination of MEP	Summary of approach(es) for the determination of MEP	Comments – Remarks
A	Wienfluss R.		MEP oriented towards the historical reference for Wienfluss	Use of historical sources to establish MEP (Wienfluss before modification) Current monitoring knowledge used. Measures are already in place giving practical information on their effectiveness	
	Bregenzerach R.			MEP is derived from natural hydrographs. Hydrological values indicative for the stress should not lay far out of the natural range	
	Danube R.			MEP is oriented to the original functioning of a natural river/floodplain system	
B	Dender R.	Macrophytes, macroinvertebrates, fish (and birds) Presence of species drawn from the ecological indicators list used in the system of stream typology	MEP = A river which meets the indicators typical for the natural condition of a large river given the presence of physical modifications, where the river based constructions do not obstruct the ecological continuum (migration), where spawn places and shelter are abundant (semi-natural banks), where erosion in the catchment area is reduced and which is influenced by natural water level fluctuations (reduced tide)	The determination of MEP is based on available information on ecological quality indicators To determine MEP, use was made of: – policy for the repair of stream types in Flanders (priority zones); – distribution and typology of streams with valuable ecology, use of ecological indicators as minimum values; – system of stream typology; and, – ecological quality objectives.	
D	Elbe R.	No information			
	Seefelder Aach R.	No information			
	Lahn R.	No information			

Table 14.27. *Continued*

Country	Name of case study	Biological elements and methods used to derive their values	Results on the determination of MEP	Summary of approach(es) for the determination of MEP	Comments – Remarks
	Ruhr R.	Macrophytes, phytoplankton, fish (elements of a lake, not a river) To derive values, expert judgement was used		MEP is derived from comparison with a comparable water body	
	Mulde R.	Definition of MEP not feasible. The ecological status of the study area is dominated by its chemical status.			
	Dhünn R.	WB not HMWB			
E	Lozoya R.	Phytoplankton (density), macrophytes (littoral, structure), benthic invertebrates, fish fauna (limited, no natural lake fish community in Spain)	Biological, hydromorphological and physicochemical quality elements were combined for defining the trophic status and fauna/flora community at MEP ⇒ MEP: Oligotrophic siliceous deep lake *Boundary values*	Steps followed: 1. Characterisation of the water body/Typology A for the next closely resembling water categories (reservoirs ⇒ lakes) 2. Reference conditions for each of the quality elements, spatially based 3. Combination of the quality elements	
SF	Kemijärvi L.	Macrophytes, macroinvertebrates, fish fauna	No definition of MEP given *No boundary values*	Steps followed: 1. Natural state before modification, biological elements 2. Effects of use, biological elements 3. Potential hydromorphological conditions (all mitigation measures) 4. Status of biological elements at MEP	
F	Authie R.		The highest value (20) of a standardised biological index (IBGN) was considered as reference for the catchment	Use of a standardised biological index (IBGN) (values 0–20) as indicator of the reference condition	MEP cannot be estimated at present. There is a need for further data (on removal of dams, opening of floodgates and fish ladder installations)

Table 14.27. *Continued*

Country	Name of case study	Biological elements and methods used to derive their values	Results on the determination of MEP	Summary of approach(es) for the determination of MEP	Comments – Remarks
	Rhone R.		No definition of MEP due to: – Lack of comparable natural water bodies of the same type (no reference found)	MEP could be approached via biological quality parameters most sensitive to hydromorphological alterations MEP was finally determined through reference to modified water bodies by means of thresholds on the basis of quality parameters from pilot site monitoring	
	Sarre R.		Minimum requirements to allow fish reference population: – Transverse constructions partially or temporarily passable – Partial or temporary communication with tributaries	MEP is represented by minimum ecological objectives set for each type of physical modification in the study area	
GR	Nestos R.	No information			
NL	Haringvliet Est.	Ecotopes (intertidal gradient), fish population/species abundance Values for ecotopes at MEP were assessed using ecological modelling. Models predicted effect of measures (new hydromorphology) on biology. Information on species at MEP was based on expert judgement and historical data	MEP: conditions with fully opened sluices of the present dam	Two approaches proposed: (a) All mitigation measures which do not have a significant adverse effect on the use are considered (b) Comparison with water body of the same type in an almost natural condition	

Table 14.27. *Continued*

Country	Name of case study	Biological elements and methods used to derive their values	Results on the determination of MEP	Summary of approach(es) for the determination of MEP	Comments – Remarks
	Hagmolenbeek-Hegebeek R.	Macroinvertebrates, fish fauna Biological quality elements according to the reference stream	*Partly boundary values*	Steps followed: 1. Determination of hydromorphology at MEP (all mitigation measures) 2. Determination of the water quality of MEP based on regional standards 3. Comparison with comparable water body 4. Prediction of the biology at MEP by expert judgement based on reference stream	
	Loosdrecht L.	Phytoplankton, zooplankton, macrophytes, fish fauna, littoral vegetation	MEP: clear mesotrophic lake dominated by macrophytes with natural banks and marshes *Boundary values*	Modelling of historical reference situation	
	Veluwerandmeren	Phytoplankton, zooplankton, macroinvertebrates, macrophytes, fish fauna, birds	MEP: clear mesotrophic lake dominated by macrophytes with natural banks and marshes *Boundary values*	Historical data on the condition before closure of the Veluwerandmeren, modelling of ecosystem functioning	
NO	Suldalslagen R.	Fish fauna, riparian vegetation	No attempt of defining boundary values at MEP	Expert judgement, comparison with Norwegian Environmental Quality Standards for Biodiversity and Nature Conservation	
	Beiarn R.	Macrophytes, macroinvertebrates	Reduced river after abstraction of water with scarce fish population and without anadromous fish *No boundary values*	Conditions with 'all mitigation measures done' stable situation assuming best practice water abstraction	

Table 14.27. *Continued*

Country	Name of case study	Biological elements and methods used to derive their values	Results on the determination of MEP	Summary of approach(es) for the determination of MEP	Comments – Remarks
S	Emån R.	Fish fauna, macroinvertebrates, macrophytes	No description *No boundary values*	'All mitigation measures' no non-physical pressures	It is not possible at this state to define the exact limit of the MEP, because too little research in the area of effects of mitigation measures have been carried out
	Dalälven R.	Fish	MEP equivalent to best possible biological qualities in spite of existing HP constructions	Main focus is on sustainability of fish populations. MEP is defined in a table of criteria types (hydromorphological, physical stress factors) and appropriate values/thresholds for both MEP and GEP.	
	Umealven R.	Fish fauna, macroinvertebrates	No definition of MEP *No boundary values*	'All mitigation measures'	
	Baltic coast	Zoobenthos as far as data is available	No definition of MEP *No boundary values*	No clear scheme or definition	
UK (EW)	Kennet R. (Thames) Tame R.		MEP of fish is defined as similar fish assemblages (densities, biomass and species diversity) to those associated *with best management practice* on similar water bodies with similar modifications	Expert opinion (partly)	In the WFD, clear qualitative classification of maximum, good and moderate potential is provided. This gives freedom for interpretation and will produce discrepancies in the classification of ecological potential due to the variability found between different experts' opinions In this case, the three qualitative categories have been translated into three quantitative categories based on deviation from the reference fish assemblages. The variations based on species composition, density and biomass

Table 14.27. *Continued*

Country	Name of case study	Biological elements and methods used to derive their values	Results on the determination of MEP	Summary of approach(es) for the determination of MEP	Comments – Remarks
	Sankey Brook Great Ouse R. (C) Main Ouse (D) Wissey				
UK (Scot)	Forth Est.	Macrophytes, benthic invertebrate fauna Fish excluded due to lack of data on certain parameters and the wide variation of population numbers from year to year which can be misleading Phytoplankton was unimportant	No change in category or type: MEP $\leq$ GES Comparisons made between MEP, GES and HES are based on an earlier draft of the support material (guidance papers) agreed by the HMWB WG for conducting the case studies Extension and area of ecotopes at MEP modelled *partly boundary values*	Process summarised in Fig. 9.1 based on an earlier draft of the support material (guidance papers) agreed by the HMWB WG for conducting the case studies. MEP is expressed in terms of the potential conditions of relevant/'most sensitive' biological quality elements derived by two independent routes which amount *to modelling procedures*. The first route predicts effects on the extant biota of eliminating the effects of non-physical pressures and applying appropriate mitigation measures. The second route examines the deviation of the biological condition of the HMWB from that of a biological reference standard (comparable WB), then adjusts to MEP	
	Tummel R.	Fish fauna, benthic invertebrate fauna (River habitat survey, RIVPACS), macrophytes and phytobenthos	– For 3 reservoirs: no change in category: MEP $\leq$ to $\ll$GES – For 4 reservoirs: change in category from river to lake: MEP $\leq$ to $\ll$ HES for lakes of the appropriate type – For 31 river stretches: no change in river type: MEP $\leq$ GES	Overall approach same as for Forth Estuary	

Table 14.27. *Continued*

Country	Name of case study	Biological elements and methods used to derive their values	Results on the determination of MEP	Summary of approach(es) for the determination of MEP	Comments – Remarks
			Comparisons made between MEP, GES and HES are based on an earlier draft of the support material (guidance papers) agreed by the HMWB WG for conducting the case studies		
	Dee R.	Fish fauna, benthic invertebrate fauna (River habitat survey), macrophytes There is lack of biological standards and the Scottish reference typology was not yet available. Therefore, targeted sampling of macroinvertebrates was carried out at sites impacted by HP as well as close unimpacted reference water bodies. A suitable *spatial network of rerence sites* was chosen. The use of the River Habitat Survey to find reference data on macroinvertebrates was also considered	*Partly boundary values* – For 1 reservoir: no change in category (lochs are lakes): MEP $\leq$ to $\ll$ GES – For 5 reservoirs: change in category from river to lake: MEP $\leq$ to $\ll$ HES for lakes of the appropriate type – For all remaining river stretches: no change in river type: MEP $\leq$ to $\ll$ GES Comparisons made between MEP, GES and HES are based on an earlier draft of the support material (guidance papers) agreed by the HMWB WG for conducting the case studies	Overall approach same as for Forth Estuary	
UK (NI)	Lagan R.	Benthic macroinvertebrates (use of RIVPACS), fish fauna Macrophytes and phytoplankton are not considered because they are mainly influenced by the artificial salinity regime		Proposed to use all three approaches, i.e. spatially-based, based on models and expert judgement RIVPACS is for example used to generate predicted invertebrate assemblages but it is not possible to estimate the proportion of differences which is a direct result of the modifications	

Table 14.28. Case study results on "closest comparable water body". *Source:* Extracted from the case studies by the authors

Country	Name of case study	Method/criteria for identification of closest comparable WB	Which is the comparable WB?	Natural or modified WB used as comparable?	Remarks
A	Wienfluss R.	Use of historical situation	Wienfluss before modification	Natural	
	Bregenzerach R.	No comparable natural WB			
	Danube R.	No comparable natural WB			
B	Dender R.	Use of other rivers in Flanders of the same large river stream type	No comparable water body found. The candidate Leie was rejected because its hydromorphological status is worse than that of the River Dender at MEP		
D	Elbe R.			Natural. The use of a modified WB was not considered as useful	
	Seefelder Aach R.			Natural. The use of a modified WB was not considered as useful	
	Lahn R.			Natural. The use of a modified WB was not considered as useful	
	Ruhr R.		Special type of lake in the lowlands (WB in question is characterised by 19.7 km stagnant water and 7.9 km free flowing river)	Only natural comparable WBs are used. In case of comparison with other HMWBs, it would be necessary to define the MEP for every water category and type depending on the use as a reference for scaling the assessment method	
	Mulde R.	Definition of MEP not feasible. The ecological status of the study area is dominated by its chemical status			
	Dhünn R.	WB not HMWB			

Table 14.28. *Continued*

Country	Name of case study	Method/criteria for identification of closest comparable WB	Which is the comparable WB?	Natural or modified WB used as comparable?	Remarks
E	Lozoya R.	Comparable water body selected with similar morphometric and altitude characteristics (Typology A) within geographic proximity Due to lack of natural water bodies (only 1), other reservoirs within the same typology class were included	Five comparable water bodies found: – Four oligotrophic siliceous high-altitude reservoirs of comparable depth. Lozoya reservoirs differ in respect to trophic status (meso to eutrophic) and water level fluctuations – One natural oligotrophic, deep glacier lake. Glacial lakes are most comparable natural water bodies to reservoirs as they have a frontal moraine, which is comparable to a dam, and moreover have fluctuating water level. Nevertheless the outlet is located at the surface as opposed to the majority of reservoirs , whose outlets are located at high or intermediate depth	Natural and modified	The closest comparable natural lakes may nevertheless still be too different from the HMWB in question, regarding reference conditions assessment and *EQR* boundaries
SF	Kemijärvi L.	No comparison with a reference lake was carried out since Lake Kemijärvi represents a rare lake type, and good data was available on the historical natural conditions		Natural	
F	Authie R.	Use of another river with similar problems	Canche river which is a coastal river with similar problems (fish migration, erosion, pollution)		
	Rhone R.	No river similar to the Rhone was found when compared to natural WBs MEP was determined through reference to modified WBs		Natural (unsuccessful) Modified (successful)	Reference system based on modified water bodies was recommended
	Sarre R.				
GR	Nestos R.	No information			

Table 14.28. *Continued*

Country	Name of case study	Method/criteria for identification of closest comparable WB	Which is the comparable WB?	Natural or modified WB used as comparable?	Remarks
NL	Haringvliet Est.	Present estuary and the one expected at MEP were compared with a set of other estuaries (modified/natural) on regional and European level. Comparison was based on parameters such as total volume, loss of intertidal area, estuary volume index	No comparable WB was found	Natural and modified	Comparable water bodies are in practice hard to find
	Hagmolenbeek-Hegebeek R.	– Same water type (middle part of a slowly flowing lowland stream) – Hydromorphology corresponding with the assumed MEP ⇒ Inventory study used (hydrology, morphology, water quality, maintenance, macro-invertebrates fish data)	One reference stream found: – Better water quality but still exceeds the standards – Biological indicators indicate good/moderate ES – Different substrate	Natural and modified	It remains risky to base MEP on the characteristics of the comparable water bodies, as two streams can be comparable to a certain extent but they are never completely the same. It would be preferable to base MEP on a reconstruction of the theoretical status using a combination of different methods (ecological prediction, models, historical data or spatial reference waters)
	Loosdrecht L.		2 reference lakes within the region which have clear state, macrophyte dominated		
	Veluwerandmeren		No comparable water body due to the reclamation of the Flevopolder		
NO	Suldalslagen R.				
	Beiarn R.	No consideration of a reference site since there is no river in the area with better documentation of the biological or physico-chemical status than Beiarelva			

Table 14.28. *Continued*

Country	Name of case study	Method/criteria for identification of closest comparable WB	Which is the comparable WB?	Natural or modified WB used as comparable?	Remarks
S	Emån R.	Identification according to stream order, morphological resemblance and anthropogenic pressures – Water body should not be impacted by non-physical pressures – Influenced by hydropower plant comparable morphology – High nature value	Part of the River Skärveteån in the Emån catchment	Modified	
	Dalälven R.	Search for water bodies with comparable fish data, mainly unregulated	No comparable WB		
	Umealven R.	Bedrock, geomorphology, regional climate, equal size and length, similar history of human use/land use	Free flowing parallel Vindel river	Natural	
	Baltic coast	No comparable water body has been identified due to very high natural variability, lack of data and lack of anthropogenic unaffected areas		(Natural)	
UK (EW)	Kennet R. (Thames)		There are a number of comparable rivers, such as the Itchen, Test and Hampshire Avon. Of these the Test also has a large number of in-channel structures while the Avon is possibly more free flowing. These rivers could provide some comparable reference conditions, but in practice, using existing methods for biotic scoring, there is little difference between them		

Table 14.28. *Continued*

Country	Name of case study	Method/criteria for identification of closest comparable WB	Which is the comparable WB?	Natural or modified WB used as comparable?	Remarks
	Tame R.		No comparable site using RHS or FDMS data could be used to define the MEP/GEP of fish for a particular stretch. The RHS data provide a quantification of the habitats *in fine*. To assess the impact of physical modifications on the fish populations in our different case studies, the RHS data were used on a broad scale (presence or absence of modifications such as embankment etc.). It was not possible to make much use of the habitat quality elements because of the way fish data are collected, without this linkage to RHS. However, for actual planning of rehabilitation works, fish ecologists will need to use RHS data on a much more precise level		
	Sankey Brook				
	Great Ouse R. (A) Main Ouse (B) Wissey				
UK (Scot)	Forth Est.	No comparable water body found (high natural variability) Analogue for the estuary was derived with a model based on the Forth itself with boundaries modified to represent the MEP condition	Forth with boundaries modified to represent the MEP condition		Derivation of an analogue water body for the Forth Estuary by modelling is expedient in that the inevitable typological discrepancies that would be introduced by using another estuary for this purpose are avoided. Moreover, this approach eliminates the need to have access to data for the analogue estuary. However, it may be helpful eventually to apply some knowledge from other estuaries

Table 14.28. *Continued*

Country	Name of case study	Method/criteria for identification of closest comparable WB	Which is the comparable WB?	Natural or modified WB used as comparable?	Remarks
	Tummel R.	The selection of reference site was map based with similar altitude, geology and as far as possible size/catchment area as the impacted sites Suite of sampling sites within the impacted and nearby unimpacted water bodies		Natural	
	Dee R.	For water bodies *with no change* in water category: assessment of MEP based on biological and hydromorphological elements compared to those of a natural analogue For water bodies *with change* in water category: assessment against the new water category natural analogues. In this case, the DHRAM method was used with reference a set of natural Scottish lochs		For water bodies *with no change* in water category: natural For water bodies *with change* in water category: natural	Comparisons with reference sites outside the Dee system were excluded due to similar data availability and compatibility issues
UK (NI)	Lagan R.	Many rivers of the same type as Lagan in Northern Ireland are degraded due to some or all of the influences found in Lagan. Hence for comparative purposes an approach using the 'best of their type' in terms of the quality of the biological elements is adopted			

Table 14.29. Case study results on the definition of GEP. *Source:* Extracted from the case studies by the authors

Country	Name of case study	Summary of approach for the determination of GEP	Mitigation measures for achieving GEP	Current ecological potential	Have Ecological Quality Ratios (*EQR*) been calculated and how?	Comments – Remarks
A	Wienfluss R.		– Ensure continuous flow – Place water-type-specific substrates in the entire course of the Wienfluss – Ensure that all weirs of the retention basis are kept open permanently and completely till the two-year flood water – Obtain a higher heterogeneity of the current and substrate conditions in the paved Wienfluss – Extend the river-bank line, vary slope inclinations, and improve river-bank structures through biology-engineering measures	Rehabilitation measures so far taken in Wienfluss show that it would be possible to attain GEP, provided that these measures are extended to the entire Wienfluss		
	Bregenzerach R.	GEP is defined as only a slight deviation from the MEP	Mitigation measures are the same as the basic measures for the POM Hydrological limits are based on realistic assumptions since threshold levels are already achieved currently and would only affect some extreme situations during the low flow winter period			

Table 14.29. *Continued*

Country	Name of case study	Summary of approach for the determination of GEP	Mitigation measures for achieving GEP	Current ecological potential	Have Ecological Quality Ratios (*EQR*) been calculated and how?	Comments – Remarks
			Mitigation measures: – Maximum daily flow amplitudes should stay within a limit of 1:5 which would probably represent an economically acceptable value. This limit would lead to a significant reduction of extreme peaks which are considered to have the most negative impacts on fish – Proposed to reduce shortterm flow alterations to a limit of 3 to 4 times the natural variation (60–80% flow alteration within 15 min). This would lead to a significant increase of the viability of the fish population			
	Danube R.	The definition of GEP is based on MEP but is influenced by the practicability of the mitigation measures. Technical feasibility and costs of mitigation measures were defined by a number of implemented or intended restoration programmes	Mitigation measures are the same as the basic measures for the POM Mitigation measures for Greifenstein: – Creating dynamic floodplains and free flowing section of the impoundment – Reconstructing the mouth of the Gießgang – Re-establishment of longitudinal and lateral connectivity			

Table 14.29. *Continued*

Country	Name of case study	Summary of approach for the determination of GEP	Mitigation measures for achieving GEP	Current ecological potential	Have Ecological Quality Ratios (*EQR*) been calculated and how?	Comments – Remarks
			– Structuring the river bottom and banks along the impoundment compatible with navigation Mitigation measures for Freudenau: – Creation of dynamic gravel bars and sand habitats in the upper part of the impoundment – Connections to riparian floodplains and lateral water bodies – Structuring of the impoundment with riparian side arms and bays – Establishment of a fish bypass channel at the weir			
B	Dender R.	Based on the definition of MEP Use of practical feasibility to distinguish between MEP and GEP	Same as for MEP without controlled reduced tide		*EQR* calculated as the ratio of *BBI* (Belgian Biotic Index) values *EQR*(BBI) = actual *BBI* / needed *BBI*	

Table 14.29. *Continued*

Country	Name of case study	Summary of approach for the determination of GEP	Mitigation measures for achieving GEP	Current ecological potential	Have Ecological Quality Ratios (*EQR*) been calculated and how?	Comments – Remarks
		GEP = A river which meets the indicators typical for the natural condition of a large river given the presence of physical modifications, where the river based constructions do not obstruct the ecological continuum (migration), where spawn places and shelter are abundant (semi-natural banks)				
D	Elbe R.	No information				
	Seefelder Aach R.	No information				
	Lahn R.	No information				
	Ruhr R.	Definition of GEP was based on deviation from the biological values at MEP	– Reducing phosphorus concentration (buffer stripes) – Restoring the stream continuum (fish migration devices at the weirs) – Ameliorating measures along the borders of the lake (introduction of reeds, removal of bank fixation on one side of the river at least)	Ecological potential considered as good (based on fish fauna assessment)		
	Mulde R.	Definition of MEP (and therefore GEP) not feasible. The ecological status of the study area is dominated by its chemical status				
	Dhünn R.	WB not HMWB				

Table 14.29. *Continued*

Country	Name of case study	Summary of approach for the determination of GEP	Mitigation measures for achieving GEP	Current ecological potential	Have Ecological Quality Ratios (*EQR*) been calculated and how?	Comments – Remarks
E	Lozoya R.	'Slight deviations': *Biological quality elements*: community of cold siliceous *mesotrophic* waters (El Atazar: oligotrophic) without macrophyte community in the littoral *Hydromorphological quality elements*: to maintain the current use, water level fluctuations would be inevitable and thus the littoral zone would remain without riparian vegetation		Mesotrophic to eutrophic (Pinilla) status of the reservoirs with anoxia in the hypolimnion GEP: Shift in trophic status: – Eutrophic to mesotrophic (Pinilla res.), – Improved mesotrophic status without anoxia (middle reservoirs) – Mesotophic to oligotrophic (El Atazar)		
SF	Kemijärvi L.	GEP and MEP as function between ecological significance and effects on use/costs (scheme)	Bottom weirs, fish stocking Various mitigation measures already in place	MEP would not differ significantly from current status GEP assumed		
F	Authie R.	The value (18/20) of the biological index (IGEN) is used as the definition of GEP	– Drastic reduction of erosion and diffuse pollution – Examination of each dam on the type of equipment needed for fish migration			
	Rhone R.	Not possible to define GEP due to the same dificulties applying in the case of MEP	Measures for reserved water flow are currently implemented in the Rhone. The results of these experiments are chances to quantify GEP on the basis of monitoring feedback			

Table 14.29. *Continued*

Country	Name of case study	Summary of approach for the determination of GEP	Mitigation measures for achieving GEP	Current ecological potential	Have Ecological Quality Ratios (*EQR*) been calculated and how?	Comments – Remarks
	Sarre R.	GEP is derived from MEP given the current technical, financial possibilities and the current use Parameter values after rehabilitation should reach a minimum level of 60% with respect to High Ecological Status (reduced to 50% when many disturbances accumulate on one section)	– Reserved flows – Fishways – Management and restoration of the banks			
GR	Nestos R.	No information				
NL	Haringvliet Est.	Practicability of measures (including costs and technical feasibility) was proposed as a way for defining GEP Biology was defined based on a model. Hydromorphology at GEP formed the basis for the new ecotopes. Values for macroinvertebrates and fish were based on expert judgement For physicochemical conditions, the generic values used in the Netherlands were applied	Measures correspond to the scenario of 'controlled tide' with ecological restoration at 80% (HES is 100% and MEP is 90%) At GEP, ecological continuum is ensured and costs are lower than for MEP			

Table 14.29. *Continued*

Country	Name of case study	Summary of approach for the determination of GEP	Mitigation measures for achieving GEP	Current ecological potential	Have Ecological Quality Ratios (*EQR*) been calculated and how?	Comments – Remarks
	Hagmolenbeek-Hegebeek R.	Application of the same definition of slight deviation between GEP and MEP as between GES and reference conditions for natural waters. GEP is one class lower than MEP	GEP may be a selection of measures for MEP taking into account the practicability and realisation Stream restoration is based on a voluntary basis and may remain incomplete (⇒ hydromophology of GEP)			
	Loosdrecht L.	Same hydromorphological and physicochemica conditions as for MEP		Eutrophic, turbid cyanobacteria dominated lake		The MEP defined above will probably lead to a clear state of Lake Loosdrecht. Abiotic values which slightly deviate from it may however lead to a turbid state of the lake in which the biological parameters deviate largely from the MEP
	Veluwerandmeren	Same hydromorphological and physicochemical conditions as for MEP		Mesotrophic, clear water lake dominated by Characeae and low ecological quality of the banks		

Table 14.29. *Continued*

Country	Name of case study	Summary of approach for the determination of GEP	Mitigation measures for achieving GEP	Current ecological potential	Have Ecological Quality Ratios (*EQR*) been calculated and how?	Comments – Remarks
NO	Suldalslagen R.		As for MEP except for: reduced erosion in tributaries, building of fish passages	Biological quality elements: lack of anadromous fish caused by artificial modification $\Rightarrow$ MEP Hydromorphological elements: upstream migration is restricted, reduced breeding and spawning grounds due to increased moss coverage and siltation $\Rightarrow$ good/moderate ecological potential Physicochemical elements: pH deviate from natural conditions, compensated by liming $\Rightarrow$ GEP Pollutants: metals, detectable > GEP		By the present mitigation activities in the water body (releases at dam, operation strategy and liming) most quality elements are at GEP or MEP. But the water body is in a transitional state where the riverbed structure could endanger the Atlantic salmon population in the long run. To avoid this, habitat improvements should be carried out
	Beiarn R.	Assuming unstable conditions below the tunnel intakes, including operation problems	There is no direct hydropower installation in the river stretches below the tunnel intakes and therefore the possibility for measures limited			
S	Emån R.	GEP is identified as the ecological status which is achieved when all mitigation measures as for MEP except fish ladders are accomplished (costs for fish ladders high)	– Minimum release of 5% – Gradually increase/decrease in tapping – Restoration of cleared and straightened stretches – Creation of peak flows			

Table 14.29. *Continued*

Country	Name of case study	Summary of approach for the determination of GEP	Mitigation measures for achieving GEP	Current ecological potential	Have Ecological Quality Ratios (*EQR*) been calculated and how?	Comments – Remarks
	Dalälven R.	Expert judgement GEP is equivalent to biological qualities enough to sustain natural fish populations in spite of existing HP constructions and with a plausible level of HP exploitation potential at the highest below 75% of today's levels	Measures are given in a table of criteria types (hydromorphological, physical stress factors) together with appropriate values/thresholds for both MEP and GEP			It is important that GEP is also considered to be good in the long run. This ought to stress the need for margins of safety when deciding both necessary restoration measures and designing informative monitoring programmes
	Umealven R.	All mitigation measures for MEP, unless they exceed the limit of 10% of production losses	Mitigation measures are the same as the basic measures for the POM			
	Baltic coast	Not discussed				
UK (EW)	Kennet R. (Thames)	There are no current scoring systems or models for linking ecological status with modifications. There needs to be a high reliance on expert opinion. The difference between MEP and GEP is somewhat arbitrary, and so a pragmatic approach has been adopted which interprets GEP as adopting best practice in river management, given the continued use of the modifications for the functions that were intended. Ecologists are then reasonably clear about what this means and can suggest mitigation measures	– Manage the existing weirs and sluices to allow as much free-flowing water as possible. This would encourage colonisation by fish, macroinvertebrates and plants which prefer shallower faster flowing water with sand and gravel substrates – Targeted structural river rehabilitation, consisting of introducing greater diversity in bank structure and long profile			

Table 14.29. *Continued*

Country	Name of case study	Summary of approach for the determination of GEP	Mitigation measures for achieving GEP	Current ecological potential	Have Ecological Quality Ratios (*EQR*) been calculated and how?	Comments – Remarks
	Tame R. Sankey Brook Great Ouse R. (A) Main Ouse (B) Wissey	Expert opinion (partly) In this case, the three qualitative categories have been translated into three quantitative categories based on deviation from the reference fish assemblages. The variations are based on species composition, density and biomass	– Creating back waters – Introduction of wing dams – Building a bypass channel – Fish ladder	Good Ecological Potential achieved		
UK (Scot)	Forth Est.	Initially, the scale from 0 to 1 is divided into 4 equal intervals corresponding to the levels of ecological potential: Maximum – *EQR* 1.00 Good – *EQR* 0.75–0.99 Moderate – *EQR* 0.50–0.74 Poor – *EQR* 0.25–0.49 Bad – *EQR* 0.00–0.24 However, this division requires verification (calibration with more biological data)		Moderate ecological potential for examined water body (WB B) (but thought to be an underestimate of true ecological potential, need for revision of *EQR* values division)	*EQR*s were derived for each selected biological element/habitat combination, and were then combined to derive an *EQR* for the WB as a whole For macrophytes, present area of saltmarsh/potential area of saltmarsh was calculated (use of modelling) For BIF, data for one area were adopted as MEP standard ($EQR = 1$) for the comparisons	The uncertainties encountered in arriving at biological interpretations of MEP and in estimating *EQR*s preclude rigorous definition of GEP levels in terms of the biological quality elements. One approach that might be explored is based on economic analysis. This would define reasonable/practical mitigation measures in terms of a cost threshold which would then be related to a minimum level of biological quality for GEP status.

Table 14.29. *Continued*

Country	Name of case study	Summary of approach for the determination of GEP	Mitigation measures for achieving GEP	Current ecological potential	Have Ecological Quality Ratios (*EQR*) been calculated and how?	Comments – Remarks
						Another possible approach would be to set a GEP threshold value for *EQR*
	Tummel R.	Initially, the scale from 0 to 1 is divided into 4 equal intervals corresponding to the levels of ecological potential: Maximum – *EQR* 1.00 Good – *EQR* 0.75–0.99 Moderate – *EQR* 0.50–0.74 Poor – *EQR* 0.25–0.49 Bad – *EQR* 0.00–0.24		Moderate to Good Ecological Potential	*EQR* calculated, based on macroinvertebrates and macrophytes for some water bodies according to data availability	It was difficult to define the *EQR* threshold value for GEP. The proposed threshold value of 0.75 is a nominal value and is likely to require adjustment in the light of more detailed assessments of the effects of various measures on the ecology
	Dee R.	Approach same as in Tummel and Forth Estuary (where *EQR* range for GEP is 0.75–0.99)			*EQR* has been calculated for lochs and rivers using the form $EQR = C / (M + C + E)$, *C* is taxa common with reference, *E* is 'exotic' taxa not recorded at reference, *M* reference taxa missing In some cases, it was not possible to say whether a low *EQR* is the effect of acidification or sediment starvation. This will be possible when a non-acidified reference is found	Inadequacy of the existing biological monitoring programme in Scotland to provide the necessary data for the definition of MEP, GEP and robust *EQR*s

Table 14.29. *Continued*

Country	Name of case study	Summary of approach for the determination of GEP	Mitigation measures for achieving GEP	Current ecological potential	Have Ecological Quality Ratios (*EQR*) been calculated and how?	Comments – Remarks
					An alternative proposed was to derive the quality ratio as the reciprocal of risk score (and calculate *EQR*s relevant to GES and *EQR*s relevant to MEP). *EQR* for MEP (physically-based *EQR*) is obtained by re-calculating the risk score to include only factors that can be removed (through 'all mitigation') without prejudice to the designated use	
UK (NI)	Lagan R.	No information				

Table 14.30. Case study results on the "Programme Of Measures" (POM). *Source:* Extracted from the case studies by the authors

Country	Name of case study	Programme of Measures (POM)		How was the link between GEP and the Programme of Measures been determined?
		Basic measures	Supplementary measures	
A	Wienfluss R.			
	Bregenzerach R.	Basic measures are identical with the mitigation measures for GEP		
	Danube R.	Basic measures are identical with the mitigation measures for GEP		
B	Dender R.	Measures for achieving GEP are seen as the basic measures because they aim at enhancing actual ecological quality. Basic measures are considered to be relevant to hydromorphological factors	Supplementary measures are more oriented to water quality: – Decrease the impact of waste water discharges – Introduction of fish species	Measures for achieving GEP are seen as the basic measures because they aim at enhancing actual ecological quality. Basic measures are considered to be relevant to hydromorphological factors
D	Elbe R.			
	Seefelder Aach R.			
	Lahn R.			
	Ruhr R.	– Reducing phosphorus concentration (buffer stripes) – Restoring the stream continuum (fish migration devices at the weirs) – Ameliorating measures along the borders of the lake (introduction of reeds, removal of bank fixation on one side of the river)		
	Mulde R.			
	Dhünn R.			

Table 14.30. *Continued*

Country	Name of case study	Programme of Measures (POM)		How was the link between GEP and the Programme of Measures been determined?
		Basic measures	Supplementary measures	
E	Lozoya R.	Measures chosen from WFD Art. 11(3) and 10: – Efficient and sustainable water use – Abstraction controls – Safeguard water quality – Controls for point source discharges and diffuse sources including creation of wetland areas, construction of wastewater treatment plants	– Central government legislative instruments: River Tajo Basin Management Plan should include programme to determine environmental quality objectives, to define and establish minimum flow, to protect ecosystems – Autonomous government legislative instruments: Resource Management for reservoirs, Forestry Plan, Director Plan of wastewater treatment – Negotiated environmental agreements among Confederación Hidrográfica del Tajo, Canal Isabel II and Autonomous Community of Madrid: for the management of reservoir operations and water withdrawal (for example, multilevel selective withdrawal). Promotion of adapted farming and agricultural production (codes of good practice) – Research, development and demonstration projects (monitoring, the regeneration of the littoral zone of reservoirs, viability of creation of wetland areas etc) – Efficiency and reuse measures, *inter alia*, promotion of a rational water use and urban water rate and water-saving techniques	
SF	Kemijärvi L.	Not relevant in the study		
F	Authie R.	– Proper function of the official Management Committee for Migrating Fish which facilitates the free movement of fish – Use of 1922 decree classifying Authie and tributaries as migrating fish rivers – Use of requirements of Natura 2000 for the area	Better agricultural practices Active information of stakeholders and support of the Water Management Plan	

Table 14.30. *Continued*

Country	Name of case study	Programme of Measures (POM)		How was the link between GEP and the Programme of Measures been determined?
		Basic measures	Supplementary measures	
	Rhone R. Sarre R.	Measures will be in line with the Rhone Action Plan and possibly Rhone Water Development and Management Plan		
GR	Nestos R.			
NL	Haringvliet Est. Hagmolenbeek-Hegebeek R.	It was not possible to distinguish between basic and supplementary measures, because the difference between the two types of measures is unclear. The chosen measures all aim to reach the environmental objectives of Article 4 and also fall under one of the instruments listed in WFD Annex VI, Part B (clear definition and criteria for the two types of measures are needed). Morphological restoration: – Development of inundation zone/creation of retention areas – Partial removal of weirs – Construction of fish passages – Passive development of meanders – Reconstruction of meanders – Removal of artificial embankment of channel – Adaptation of channel maintaining the capacity – Application of holes and sandbanks – Application of disturbing objects – Application of structure adapted to target species – Creation of steep and overhanging banks – Application of pools Hydrological restoration: – Restoration of the former catchment area		

Table 14.30. *Continued*

Country	Name of case study	Programme of Measures (POM)		How was the link between GEP and the Programme of Measures been determined?
		Basic measures	Supplementary measures	
		Water quality measures: – Reduction of manure emissions due to national policy – Reduction of household emissions due to national policy – Reduction of pesticide emissions due to national policy – Reduction of other diffuse pollution due to national policy – Reduction of diffuse pollution from two stormwater overflows Other measures: – Ecological maintenance of stream channel – Ecological maintenance of banks – Re-introduction of species		
	Loosdrecht L.	The measures for achieving GEP are the same as for MEP except for the restoration of the banks: – Decrease the turbidity caused by the resuspended silt – Possible measures to reduce the phosphorus load into Loosdrecht are: 1. Reducing the external loading via the inlet of allochtonous water by removing the phosphate in the water in a chemical way (dephosporization) or by reducing the inlet of the relatively phosphate rich seepage water from the Bethune polder 2. Reducing the external loading via diffuse emissions of agriculture, recreation and spill-overs in the catchment 3. Reducing the internal loading by biomanipulation (Removal of a large fish biomass)		

Table 14.30. *Continued*

Country	Name of case study	Programme of Measures (POM)		How was the link between GEP and the Programme of Measures been determined?
		Basic measures	Supplementary measures	
	Veluwerandmeren	Future measures and actions for the Veluwerandmeren are based on an integrated plan made by the authorities and parties concerned (present water users) in an open plan process. The following measures aim at an improvement of the ecological function of the lake: – Realisation of research on natural water level dynamics and the economic and ecological costs and benefits – Improvement of sewage works – Creation of collection point for waste water from yachts – Development of a reed area as habitat for birds, amphibians and fish – Limitation of recreation activities in important breeding area for birds in the Veluwemeer – Marsh development in Nuldernauw to improve the migration possibilities of fish, amphibians and mammals – Construction of fish passage for eels at the Roggebotssluice to improve migration into the Veluwemeer – Marsh development in Wolderwijd (southern of Harderwijk) – Realisation of 5 km natural banks near Elburg by displacement of recreational activities and houses		
NO	Suldalslagen R. Beiarn F.			
S	Emån R.	– Assessment of impact of water withdrawals – Costs for water services gathered in trust funds to finance ecosystem restoration, public education – Minimum tapping to protect species under the Birds Directive	– Environmental certificates for hydropower plants linked to WFD requirements for MEP/GEP – Demonstration and research projects about river restoration, in order to inform the public and investigate the effects of river restoration – Time-limited licences for hydropower plants – Water-saving measures for industries and agriculture	
	Dalälven R.	– CEN methodology for monitoring programmes	– Specific variables or indexes would be especially useful for measuring effects of physical modification. Such measures should be able to extract e.g. indicator species from data collected following CEN standards	

Table 14.30. *Continued*

Country	Name of case study	Programme of Measures (POM)		How was the link between GEP and the Programme of Measures been determined?
		Basic measures	Supplementary measures	
			– The frequency of monitoring could be increased over the season to be able to more accurately obtain information from critical periods, for example during migration of fish. Increase in frequency is of course on condition that the monitoring will not affect the populations to any measurable degree due to increased mortality	
	Umealven R.	All measures suggested to achieve GEP are considered as 'basic'	No need for supplementary measures since all measures necessary to achieve GEP are considered as basic, and the measures deemed too costly are not implemented at all	All measures suggested to achieve GEP are considered as 'basic' All measures undertaken to achieve GEP have to be carefully monitored to ensure that they meet the expectations. Such monitoring programmes should, however, be individually designed for each measure or combination of measures
	Baltic coast			
UK (EW)	Kennet R. (Thames)			
	Tame R.			
	Sankey Brook			
	Great Ouse R. (A) Main Ouse (B) Wissey	(A) Ensure best possible status in the parallel bypass channels		
UK (Scot)	Forth Est.	Measures related to the Directives on waste water treatment, nutrients and pollution, the Habitats Directive and the Birds Directive	– Provision of agricultural compensation for managed retreat – Collaboration with electricity operators for code of practice on environmental impact minimisation from abstraction – Return of cooling water	

Table 14.30. *Continued*

Country	Name of case study	Programme of Measures (POM)		How was the link between GEP and the Programme of Measures been determined?
		Basic measures	Supplementary measures	
	Tummel R.	A selection of the proposed basic measures is: – Sediment-starvation downstream of intake structures should be reversed – Compensation flow provision at all abstraction points where the natural catchment area exceeds 20 km^2 and where normal flows are wholly diverted by hydro structures – Provisions for fish passage should be made if a hydro structure is presently known to prevent natural migration and if any measures exist which are technically feasible for the site in question – Stratification effects at power stations should be reversed	A selection of the proposed supplementary measures is: – Control of water levels to achieve ecological targets in terms of the timing and magnitudes of extreme high and low water levels – Compensation flow and freshet releases should be reviewed if evidence suggests that existing quantities of water released for environmental benefit could be used to deliver greater ecological benefit in the future – Phasing of generation at power stations where evidence indicates that the abrupt start and end to power generation is causing ecological problems	Measures for the achievement of GEP were proposed on the basis of the assumption that any measure currently responsible for downgrading of ecological status should be the subject of reasonable measures for improvement so long as these do not significantly impact adversely on hydropower generation. Links between GEP and measures are therefore based on the present understanding of physical factors which are known or thought to exercise control on the ecological status
	Dee R.	Catchwater intakes: periodic provision of sediments by developing practicable codes of practice, provisions for fish passages Impounding dams: improve the efficiency of existing fish passages, provision of fish passages where no exist, compensation flows, periodic provision of sediment Power stations: remediation of thermal effects, alteration of generation discharges Control weirs: periodic provision of sediment, alteration of flow releases	Impounding dams: control of water levels in terms of timing and magnitude of extreme high and low water level, compensation flows Power stations: phasing of power generation	Basic measures are proposed for the achievement of GEP on the basis of the assumption that any measure currently responsible for downgrading of ecological status should be the subject of reasonable measures for improvement so long as these do not significantly impact adversely on hydropower generation. Links between GEP and measures are therefore based on present understanding of physical factors which are known or thought to exercise controls on ecological status
UK (NI)	Lagan R.			